Explanation

Explanation

Papers and discussions by:
PETER ACHINSTEIN
RENFORD BAMBROUGH
L.JONATHAN COHEN
PETER GEACH
R. HARRÉ
GRETE HENRY
MARY HESSE
MARTIN HOLLIS
J. L. MACKIE
D. H. MELLOR
WESLEY C. SALMON
PETER WINCH

Edited by
STEPHAN KÖRNER

New Haven · Yale University Press · 1975

Library of Congress catalogue card number: 74-21348
International standard book number: 0-300-01827-4

Printed and bound in Great Britain by
The Camelot Press Ltd, Southampton

Contents

Preface   vii

I The Object of Explanation
PETER ACHINSTEIN   I
    Comment: MARY HESSE   45
    Comment: R. HARRÉ   54
    Reply: PETER ACHINSTEIN   66

II Teleological Explanation
PETER GEACH   76
    Comment: PETER WINCH   95
    Comment: GRETE HENRY   105
    Reply: PETER GEACH   112

III Theoretical Explanation
WESLEY C. SALMON   118
    Comment: D. H. MELLOR   146
    Comment: L. JONATHAN COHEN   152
    Reply: WESLEY C. SALMON   160

IV Ideological Explanation
J. L. MACKIE   185
    Comment: RENFORD BAMBROUGH   197
    Comment: MARTIN HOLLIS   205
    Reply: J. L. MACKIE   214

Index   217

Contents

Preface   vii

I The Object of Explanation
PETER ACHINSTEIN   I
  Comment: MARY HESSE   45
  Comment: R. HARRÉ   54
  Reply: PETER ACHINSTEIN   66

II Teleological Explanation
PETER GEACH   76
  Comment: PETER WINCH   95
  Comment: GRETE HENRY   105
  Reply: PETER GEACH   112

III Theoretical Explanation
WESLEY C. SALMON   118
  Comment: D. H. MELLOR   146
  Comment: L. JONATHAN COHEN   152
  Reply: WESLEY C. SALMON   160

IV Ideological Explanation
J. L. MACKIE   185
  Comment: RENFORD BAMBROUGH   197
  Comment: MARTIN HOLLIS   205
  Reply: J. L. MACKIE   214

Index   217

Preface

The papers and discussions contained in this volume are devoted to closely interrelated aspects of explanation within, and outside, the sciences. The contributors, all of whom are well known for their philosophical works, have not been content to restate their positions but have developed them further, in particular by reacting to objections raised in the discussion. In this respect the volume resembles its predecessor, which was devoted to the problem of practical reason. The essays and discussions of the present volume examine in some detail, and in new ways, the nature of scientific explanation in general, the rôle of statistics in the natural and social sciences and the function of teleology and ideology. The topics have been selected both for their intrinsic philosophical interest and for the persistence with which they confront not only the philosopher of science but also the scientist. This applies in particular to the rôle of statistics in physics, which has been in the forefront of interest ever since quantum mechanics emerged as a purely statistical theory, and to the explanatory function of teleology, which has, once again, become important to the social sciences.

I should like to thank the contributors for their contributions and my colleagues, Messrs. Michael Welbourne and David Hirschmann, for their help with the organization of the Conference and the preparation of this volume.

January 1974   S. K.

I/The Object of Explanation

Peter Achinstein
The Johns Hopkins University

I. INTRODUCTION

It is tempting to suppose that sentences such as 'Plato explained why Socrates died' or 'kinetic theory explains the gas laws' are used to express a relationship between an explainer and some object which is explained. Hereafter sentences such as these that mention an explainer and something explained will be called explanatory sentences. They may contain clauses modifying the explainer ('Plato, Socrates' student, explained why Socrates died') or the verb of explanation ('Plato carefully explained why Socrates died'). My concern here is not with the identity of the explainer—for example, it might be a person, a theory, or something else—but with the identity of the object.

The expression following the explanatory verb may or may not purport to refer to an object of explanation. For example, on one view, the explanatory sentence 'Plato explained why Socrates died' is used to express a relationship between Plato and the *event* of Socrates' death—it is this event that Plato explained—even though 'why Socrates died' is not an expression purporting to refer to this event. We might distinguish, then, between *restructured* and *unrestructured* occurrences of 'explains' as those in which 'explains' is (not) followed by a term or expression purporting to refer to an object of explanation. For example, on the event view, the occurrence of 'explained' in

1 Plato explained why Socrates died

is unrestructured while its occurrence in

2 Plato explained (the event of) Socrates' death

is restructured. We can also speak of an explanatory *sentence* as being restructured (unrestructured) if the occurrence of 'explains' in it is restructured (unrestructured).

Views according to which an explanatory sentence is used to express a relationship between an explainer and some object of explanation will be called object-of-explanation views. They are committed to the following assumption:

I   When a speaker uses an explanatory sentence S_1 there is some explanatory sentence S_2 (there may be more than one) with a restructured occurrence of 'explains' that expresses what the speaker means to be saying and therefore has the same truth-value as S_1.

We might say that any explanatory sentence is *paraphrasable* into a restructured one. For example, on the event view, sentence (1) is paraphrasable into sentence (2). And, according to object-of-explanation views, if S_2 is a restructured paraphrase of unrestructured S_1, then the object to which the expression following 'explains' in S_2 purports to refer is an object of explanation in S_1.

Object-of-explanation views are also committed to the following assumption:

II   In restructured explanatory sentences terms purporting to refer to objects of explanation and to explainers occur *purely referentially.*

This assumption might be better expressed by dividing it into two sub-theses.

IIA From a restructured explanatory sentence one may infer sentences of the form '($\exists x$) (. . . explains x) and '($\exists x$) (x explains . . .)'.

IIB Given two restructured explanatory sentences S_1 and S_2, if S_2 can be obtained from S_1 by substituting a co-referring term for the explainer term or for the object term, then S_1 and S_2 have the same truth-value.

For example, on an event view assumption IIA requires that from (2) we may infer

3 ($\exists x$) (Plato explained (event) x).

If there was no such event then (2) is false. On an event view, if 'Plato' and 'the author of the *Republic*' refer to the same person and 'Socrates' death' and 'Xantippe's husband's death' refer to the same event, then assumption IIB requires that (2) and

4 The author of the *Republic* explained Xantippe's husband's death

have the same truth-value.

What is being said about an object when it is identified as an object of explanation? First, at least, that it is what is explained. To say that in (1) the event of Socrates' death is an object of explanation is to say, at least, that Plato explained this event. More importantly, to say that certain objects but not others are objects of explanation is to make the epistemological claim that certain objects, but not others, are such that a knowledge of them puts us in a position to seek an explanans. On the event view, e.g., we are in such a position if we have identified an event, but we are not if we have identified only an individual. According to this view, we are in a position to seek an explanans if we identify what is to be explained as the event of Socrates' death, we are not if we identify what is to be explained simply as Socrates. Finally, to speak of certain objects but not others as objects of explanation is to be committed to the view that general theories of explanation need be developed only for those objects. For example, it might be held that a theory of explanation is to be provided by analysing or supplying truth-conditions for restructured explanatory sentences only. Thus, the standard Deductive-Nomological model of explanation provides truth-conditions for restructured explanatory sentences in which the expression following 'explains' designates a sentence. Once such analyses or truth-conditions are provided for the restructured sentences they are also provided for the unrestructured sentences, by assumption I.

Various types of object-of-explanation views are possible. It might be held, e.g., that any explanatory sentence whatever is restructured if its object term refers to an object of a certain type and that every unrestructured explanatory sentence is paraphrasable into a restructured one. In the next three sections I shall consider views more limited than this. These views are committed to theses of one of the following sorts:

A   Any explanatory sentence of kind K is paraphrasable into a restructured explanatory sentence whose object term refers to an object of type O. And any explanatory sentence of kind K is restructured only if its object term refers to an object of type O.

B   Any explanatory sentence of kind K is paraphrasable into a restructured explanatory sentence whose object term refers to an object of type O.

One who holds either of these types of views will admit that an explanatory sentence of kind K may have more than one restructured paraphrase. He might also hold that explanatory sentences of different kinds have different kinds of restructured paraphrases. Accordingly, such a person might say that

2 Plato explained Socrates' death

and

5 Albert explained the first sentence of the *Declaration of Independence*

are both restructured even though the object term in (2)

refers to an event and that in (5) refers to a sentence.[1] The difference between *A* and *B* is that one who holds *B* but not *A* can go on to say that an explanatory sentence of kind *K* can have one restructured paraphrase in which the object term refers to an object of type *O* and another in which it refers to an object of a different type. For example, he might hold that both (2) and

6 Plato explained the sentence 'Socrates died'

are restructured paraphrases of (1), i.e. that both an event and a sentence can be taken to be objects of explanation in (1).[2]

The views to be considered in the next three sections are all concerned with explanatory sentences of the same kind: ones in which are cited such non-linguistic items as events, states of affairs, phenomena, facts, conditions, processes, and the like. In these sentences the expression following 'explains' is either a singular term purporting to refer to an event (state of affairs, etc.) or else an expression containing an interrogative term or phrase followed by a sentence describing an event, etc. Examples of such sentences are (1), (2) and (4) above. They are the kind of explanatory sentences to which philosophers traditionally have devoted most of their attention in developing theories of explanation. For this reason I shall refer to them as *preferred* explanatory sentences.

In what follows I shall examine a number of object-of-explanation views dealing with preferred explanatory sentences,

---

[1] Such a person might also claim that the meaning of 'explain' can vary depending on the type of object being explained. For example, he might say that when we speak of an event as what is explained we do not mean by 'explain' what we do when we speak of a sentence as being explained; i.e. 'explained' in (2) and (5) has different senses. In what follows I shall assume that the meaning of 'explain' does not change with a change in object type (or with a change in explainer). (For an analysis of explanation that can be used to defend such an assumption see my *Law and Explanation*, Oxford, 1971, ch. 4.) However, the points that I regard as most important in this paper could be made without this assumption. One could hold an object-of-explanation view and still say that 'explained' in (2) and (5) has different meanings.

[2] One who said this would, of course, have to hold that 'explained' does not have different senses in (2) and (6).

as well as a view which denies that explanatory sentences express a relationship between an explainer and an object. At least some of the problems generated by traditional accounts of explanation arise from faulty ideas about the object of explanation.

2   THE OBJECT AS A NON-LINGUISTIC ENTITY

It seems commonsensical to say that among the items explained are events, facts, states of affairs, or phenomena in the real world. Indeed even some of those philosophers who finally adopt what seems to be a quite different position find it difficult to avoid speaking this way at least at the outset. Ernest Nagel after citing three examples of explanations writes:

> In both the second and third examples, the explanandum is a historical fact. However, in the second the fact is an individual event, while in the third it is a statistical phenomenon. [3]

And Hempel and Oppenheim, although they construe the explanandum as a sentence, do suggest that what is to be explained is not the sentence but the phenomenon which that sentence describes:

> By the explanandum we understand the sentence describing the *phenomenon* to be explained. [4]

In such passages these authors are concerned with preferred explanatory sentences. The first object-of-explanation view that I want to consider, then, is

> *The Non-Linguistic View:* Any preferred explanatory sentence is paraphrasable into a restructured one whose object term refers to an event, state of affairs, phenomenon, fact, etc. (And any preferred explanatory sentence is restructured only if its object term refers to an event, etc.)

[3] Ernest Nagel, *The Structure of Science* (N.Y., 1961), p. 21.
[4] Carl G. Hempel, *Aspects of Scientific Explanation* (N.Y., 1965), p. 247; my emphasis.

On either version of this view—on the version with the parenthesized material or without it—the preferred explanatory
sentence

1 Plato explained why Socrates died

has as a restructured paraphrase

2 Plato explained Socrates' death.[5]

There are two major problems with such a view (in either
version). One is a problem shared by each of the other views to
be discussed in sections 3 and 4 and I shall wait until section 5
to consider it. The other can be presented by considering the
explanatory sentences (1) and

3 Plato explained why Socrates died from drinking hemlock,

which on the present view, let us suppose, are paraphrasable,
respectively, into the restructured explanatory sentences (2)
and

4 Plato explained Socrates' death from drinking hemlock.

Now, let us assume, (1) is true but (3) is false, at least on one
plausible reading of these sentences. (Later we shall consider
the question of various readings of explanatory sentences due
to differences in emphasis.) Although Plato explained why
Socrates died he did not explain why Socrates died from
drinking hemlock. (For example he did not explain why the
hemlock made Socrates die or why hemlock was used.) Accordingly, if the paraphrases are satisfactory, (2) must be true but
(4) false. Yet the event being referred to by 'Socrates' death' in
(2) is the same event as that being referred to by 'Socrates'
death from drinking hemlock' in (4). We have here one event

[5] One who supports either version of the non-linguistic view will
agree that (2) (or at least some such sentence in which the object term
refers to a non-linguistic item) is a restructured paraphrase of (1),
though a non-linguistic theorist who rejects the parenthesized material
above will assert that in addition there may be other restructured
paraphrases in which the object term does not refer to a non-linguistic
item.

not two, an event which can be referred to by either expression. On the present view, then, (4) can be obtained from (2) and *vice versa* by substituting a co-referring term for the object-term, and therefore assumption IIB requires that they have the same truth-value, contrary to what we have been saying. This means that either the present view cannot adequately paraphrase (1) and (3) into restructured explanatory sentences about events, so that assumption I is violated for events as objects of explanation, or else that the paraphrases of (1) and (3) into (2) and (4) are satisfactory and assumption IIB is violated.

Presupposed by this argument, of course, is a view about the identity of events. We are supposing that Socrates' death is the same event as Socrates' death from drinking hemlock. And such a view, although defended by Davidson,[6] has been challenged by Kim[7] and Goldman.[8] The latter say that events are things having properties at a time and that events are identical iff the things, properties, and times are identical. But in the above example although the things and times are identical the properties are not. The property of having died is not identical with the property of having died from drinking hemlock, since, for one thing, the predicates expressing these properties aren't even coextensive.

I believe that there is a confusion here and that Davidson is talking about one sort of thing and Kim and Goldman about another. There is the event of Socrates' death, which, as Davidson urges, can be variously described as Socrates' death or as Socrates' death from drinking hemlock. But there is also what might be called the *state of affairs* which consists of Socrates' having the property of having died, and this is different from the state of affairs of Socrates' having the pro-

[6] Donald Davidson, 'The Individuation of Events', *Essays in Honour of Carl G. Hempel*, ed. Nicholas Rescher (Dordrecht, 1969), pp. 216–34.

[7] Jaegwon Kim, 'On the Psycho-Physical Identity Theory', *American Philosophical Quarterly*, 3 (1966), pp. 227–35.

[8] Alvin I. Goldman, *A Theory of Human Action* (Englewood Cliffs, N.J., 1970).

perty of having died from drinking hemlock (just as we might say that the fact that Socrates died and the fact that he died from drinking hemlock are different, though related, *facts* about Socrates). Socrates' death (the event) is something that occurred rather quickly and painlessly and was witnessed by Phaedo, his having the property of having died (the state of affairs) is none of these things. The identity criteria of Kim and Goldman are criteria for states of affairs not for events.

A defender of the non-linguistic view might now reply by agreeing that events (construed à la Davidson) cannot be objects of explanation, but he might urge that other non-linguistic entities such as states of affairs (construed à la Kim) can be. He might claim that (1) and (3) can be paraphrased not into (2) and (4) but into

5   Plato explained (the state of affairs of) Socrates' having the property of having died
6   Plato explained (the state of affairs of) Socrates' having the property of having died from drinking hemlock.

Since the states of affairs cited in (5) and (6) are not identical the previous difficulty does not arise.

Although this approach works with the present example there are other examples that prove troublesome. Consider the state of affairs of this gas' having the property of having an absolute temperature of 300°K. According to statistical mechanics this state of affairs can be explained by appeal to the state of affairs of this gas' having a mean molecular kinetic energy of $6 \cdot 21 \times 10^{-21}$ joule (or, if you like, an appeal to the latter would be part of an explanation of the former). But now we run into difficulty, because the state of affairs of this gas' having an absolute temperature of 300°K is, according to many, identical with the state of affairs of this gas' having a mean molecular kinetic energy of $6 \cdot 21 \times 10^{-21}$ joule. Both descriptions describe the same state of affairs, the former being a macro-description, the latter a micro-description. But if a restructured explanatory sentence can relate an explainer and a state of affairs, no matter how the latter is referred to, then the state of affairs of this gas' having a mean molecular kinetic energy of $6 \cdot 21 \times 10^{-21}$ joule can be explained by appeal to the state of affairs of this gas' having a mean molecular kinetic energy of $6 \cdot 21 \times 10^{-21}$ joule. But this is blatant self-explanation and must be rejected. (Even if one holds a sentential view of the explainer presumably one wouldn't want to say that the state of affairs of this gas' having a mean molecular kinetic energy of $6 \cdot 21 \times 10^{-21}$ joule is explained by an explanans containing the sentence 'this gas has a mean molecular kinetic energy of $6 \cdot 21 \times 10^{-21}$ joule'.) Rather than denying the truth of the theoretical identity the more plausible line would be to admit that the explanatory sentence in question does not relate an explainer and the state of affairs of this gas' having an absolute temperature of 300°K.

3   THE OBJECT AS LINGUISTIC

The most widespread philosophical view is that a restructured explanatory sentence relates an explainer and a sentence (or perhaps a statement). As previously noted, according to the standard D–N theory of explanation of Hempel and Oppenheim the explanandum is a sentence which gets explained by being deduced from a set of sentences satisfying certain conditions. Israel Scheffler, after considering various ways to construe the object of explanation, ends up by treating it as a sentence.[9] And Donald Davidson, who holds that in singular causal statements the causal relation relates events not sentences, has a different view about explanation. 'Explanations,' he writes, 'typically relate statements, not events.'[10]

These authors, it seems fair to say, would support

The Linguistic View: Any preferred explanatory sentence is paraphrasable into a restructured one whose object term

[9] Israel Scheffler, *The Anatomy of Inquiry* (N.Y., 1963), p. 72.
[10] Donald Davidson, 'Causal Relations,' *Journal of Philosophy*, LXIV (1967), p. 703.

refers to a sentence. (And any preferred explanatory sentence is restructured only if its object term refers to a sentence.)

On this view, for example, the preferred explanatory sentence

1   Plato explained why Socrates died

has as a restructured paraphrase

2   Plato explained the sentence 'Socrates died'.

Hempel and Oppenheim, indeed, would probably want to extend the class of preferred explanatory sentences to include explanatory sentences citing *laws* (which they construe as sentences of a certain kind). On this view, for example, the preferred explanatory sentence

3   Newton explained why the law of conservation of momentum holds

has as a restructured paraphrase

4   Newton explained the law of conservation of momentum.

The sentential view, like the non-linguistic view, is beset by a major problem to be examined in section 5. For the moment one reservation will be noted. The sentential view fails to expose the validity of certain kinds of inferences from explanatory sentences. Earlier we saw that on the non-linguistic view either unrestructured explanatory sentences are not adequately paraphrased into restructured ones about events, or else terms that purport to refer to events as objects of explanation do not occur purely referentially (i.e. restructured explanatory sentences about events are referentially opaque in the object position). If the latter alternative is accepted then, on the non-linguistic view, we cannot infer 'Plato explained Socrates' death from drinking hemlock' from 'Plato explained Socrates' death' and 'Socrates' death = Socrates' death from drinking hemlock'. If this alternative is chosen we might say that explanatory sentences of the form '*A* explained *x*'s *F* (Socrates' death)' are not *event-transparent*, meaning that if (the event of) *x*'s *F* = (the event of) *y*'s *G*, from '*A* explained *x*'s *F*' we cannot infer '*A* explained *y*'s *G*'. Whatever our views about the validity of such inferences, there is another type of inference that is, I think, less controversial. From 'Plato explained Socrates' death' and 'Socrates = Xantippe's husband' we can infer 'Plato explained Xantippe's husband's death'. Or at least there is a use of 'explain'—characterized by what I shall call *thing-transparency*—that sanctions such an inference. A sentence of the form '*A* explained . . . *x* . . .' (where variable *x* ranges over things) is thing-transparent if the *x*-position is transparent. Now if the object of explanation is construed as a *sentence* then the justification for inferences generated by thing-transparency remains unclarified. In general, from '*A* bears *R* to the sentence ". . . *x* . . ."' and '*x* = *y*' we cannot infer '*A* bears *R* to the sentence ". . . *y* . . ."'.

4   THE OBJECT AS A NON-LINGUISTIC ENTITY UNDER A DESCRIPTION

It might be said that a restructured explanatory sentence does not relate an explainer and an event *simpliciter*,[11] nor an explainer and a sentence, but rather an explainer and an event *under a description*. In explaining a human action by giving what Davidson calls the primary reason, what is explained says Davidson, is neither the action nor some description of it but rather the action under some description.[12] On this view, to explain Socrates' death under the description 'Socrates died' (or 'Socrates' death') is not to explain the same thing as Socrates' death under the different description 'Socrates died from drinking hemlock' (or 'Socrates' death from drinking hemlock').

If an event under a description is neither an event nor a description what kind of thing is it? One proposal is to treat it as an ordered pair one of whose members is an event and the

[11] In what follows I shall use the term 'event' to cover all of the types of non-linguistic objects alluded to in section 2.

[12] Donald Davidson, 'Actions, Reasons, and Causes', *Journal of Philosophy*, LX (1963), p. 687.

other a description of it, for the present let us say a sentence. We can then formulate

The Ordered Pair View: Any preferred explanatory sentence is paraphrasable into a restructured one in which the object term refers to an ordered pair consisting of an event, etc., and a sentence describing it. (And any preferred explanatory sentence is restructured only if its object term refers to such an ordered pair.)

For example, as a restructured paraphrase of 'Plato explained why Socrates died' we might write

1   Plato explained (Socrates' death, 'Socrates died')

in which the first member of the ordered pair is an event, viz. Socrates' death, and the second is a description of that event, viz. the sentence 'Socrates died'.[13] Both positions in the ordered pair are referentially transparent, so if 'Xantippe's husband's death' refers to the same event as does 'Socrates' death', and if *S* stands for the sentence 'Socrates died', then if (1) is true so is

2   Plato explained (Xantippe's husband's death, *S*).

However, what we are not permitted to do is substitute different descriptions of the event in question for the second member of the ordered pair. Even if 'Socrates died' and 'Socrates died from drinking hemlock' describe the same event, we are not permitted to go from (1) to

3   Plato explained (Socrates' death, 'Socrates died from drinking hemlock').

Furthermore, although we are permitted to go from (1) and

[13] We could just as easily take the description to be a phrase such as 'Socrates' death.' This would accord with Davidson's view that it is this phrase and not the sentence 'Socrates died' that refers to Socrates' death. This view will not here be discussed, but it should be noted that criticisms of the ordered pair view to be developed above apply *mutatis mutandis* to an ordered pair view in which the description is a phrase.

'(the event of) Socrates' death = (the event of) Socrates' death from drinking hemlock' to

4   Plato explained (Socrates' death from drinking hemlock, 'Socrates died'),

we are not permitted to go from (4) to

5   Plato explained (Socrates' death from drinking hemlock, 'Socrates died from drinking hemlock'),

even though 'Socrates died from drinking hemlock' describes the event of Socrates' death from drinking hemlock. The latter inference is not permitted since the description in (5) is not identical with the description in (4) or in (1), although it describes the same event as in (4) and (1).

Treating an ordered pair but not an event or a description as an object of explanation in preferred explanatory sentences helps us to see in what sense 'self-explanations' are and are not possible. Earlier I spoke of explaining the state of affairs of this gas' having an absolute temperature of 300°K by appeal to its having a mean molecular kinetic energy of $6 \cdot 21 \times 10^{-21}$ joule, even though these states are identical. And I said that we cannot explain the state of affairs of this gas' having a mean molecular kinetic energy of $6 \cdot 21 \times 10^{-21}$ joule by appeal to its having a mean molecular kinetic energy of $6 \cdot 21 \times 10^{-21}$ joule. If we take the object of explanation to be an ordered pair, then this difference can be expressed as follows: Where A is the ordered pair (this gas' having a mean molecular kinetic energy of $6 \cdot 21 \times 10^{-21}$ joule, 'this gas has a mean molecular kinetic energy of $6 \cdot 21 \times 10^{-21}$ joule'), and B is the ordered pair (this gas' having a mean molecular kinetic energy of $6 \cdot 21 \times 10^{-21}$ joule, 'this gas has a temperature of 300°K'), we can cite the ordered pair A in explaining the ordered pair B, but we cannot cite the ordered pair A in explaining the ordered pair A. More generally, if we have a state of affairs S and two different descriptions of it d_1 and d_2, it can be the case that

We explain (S, d_2) by citing (S, d_1) (or by citing just d_1)

but it cannot be the case that

We explain (S, d_1) by citing (S, d_1) (or just d_1).

5 A MAJOR PROBLEM

Even if the ordered pair view is superior to the others, there is a major flaw in it as well as in the others I have mentioned. The problem is that one of the basic assumptions of object-of-explanation views, assumption I, is not satisfied by any of these views. Consider the explanatory sentences

1 Smith explained why Kissinger warmly shook hands with Le Duc Tho
2 Smith explained how Kissinger warmly shook hands with Le Duc Tho,

and let us suppose that (1) is true but (2) is false. On each of the views discussed the same object is an object of explanation for both (1) and (2). According to the non-linguistic view Kissinger's warmly shaking hands, etc. (an event, let us suppose) is an object of explanation for both (1) and (2). On the linguistic view the sentence 'Kissinger warmly shook hands, etc.' is such an object. And on the ordered pair view the ordered pair consisting of the event of Kissinger's warmly shaking hands, etc., and the sentence 'Kissinger warmly shook hands, etc.' is an object for (1) and (2). That is, on these views, respectively, (1) and (2) are both paraphrasable into the restructured sentences

> Smith explained the event of Kissinger's warmly shaking hands, etc.
> (the non-linguistic view)
> Smith explained the sentence 'Kissinger warmly shook hands, etc.'
> (the linguistic view)
> Smith explained (Kissinger's warmly shaking hands, etc., 'Kissinger warmly shook hands, etc.')
> (the ordered pair view)

Since on each of these views there are identical restructured paraphrases of (1) and (2), these paraphrases will have the same truth-value. But according to assumption I, if (1) and (2) have different truth-values then so must their restructured paraphrases. So assumption I is violated by these examples, on the views considered.

The problem can also be put in the following way. When we have identified an event, a description, or an event under a description, we have not yet identified what is to be explained, since different questions can be raised each calling for a different explanans. The usual theories of explanation equate explaining with explaining *why*, but this is only one of several interrogatives possible. With respect to the above event and description we may explain *how* Kissinger warmly shook hands (suppose, for example Kissinger's arms on that day had literally been twisted). We might even explain *where* he warmly shook hands. (Someone might be puzzled about this because he thinks he knows all the places where Kissinger might have shaken hands with Le Duc Tho, yet all of these are precluded by other views he holds. To such a person we could explain where Kissinger warmly shook hands by indicating the correct answer to the appropriate question and by showing the person why this answer is consonant with his other views, or by giving him information that will change his other views so that this answer becomes consonant with them.)[14] And to explain *how* Kissinger did what he did, or *where* he did it, or *why* he did it, is to explain something different in each case.

It might be thought that this problem could be solved by restricting explanation to explanation-why. A question, then, *would* be uniquely determined by the event and description, so that being able to identify the latter would put us in a position to know what question is being raised. Besides unduly restricting the concept of explanation (explanation, even in the sciences, is not limited to explanation-why), such a view fails since a unique question is not thereby determined. Consider again the event described by the sentence

[14] See my *Law and Explanation*, pp. 73, 76–8.

S: Kissinger warmly shook hands with Le Duc Tho.

Even if we restrict explanation to explanation-why, different questions are possible with respect to the event under description *S*, each calling for a different explanans. Thus,

3   Why did *Kissinger* (rather than, say, General Haig) warmly shake hands with Le Duc Tho?

4   Why did Kissinger warmly shake hands with *Le Duc Tho*?

5   Why did Kissinger *warmly* shake hands with Le Duc Tho?[15]

That these are different questions is reflected in the fact that they have different (correct) answers. (3) might be answered by saying that Kissinger was the principal American negotiator (and General Haig was not), (4) by saying that Le Duc Tho was the principal North Vietnamese negotiator, and (5) by saying that Kissinger wanted newsmen to believe that the talks were going well. Having identified the event under description *S* we do not thereby identify an object of explanation since we do not yet know which of questions (3) to (5) or others is to be answered, even if we restrict explanation to why-questions.

Furthermore, assumption I is again violated, for consider the sentences

6   Smith explained why *Kissinger* warmly shook hands with Le Duc Tho

7   Smith explained why Kissinger warmly shook hands with *Le Duc Tho*,

and let us suppose that (6) is true but (7) is not. That is, we are supposing that Smith explained why it was Kissinger who performed the act, but he did not explain why it was with Le Duc Tho that Kissinger performed the act. On each of the views so far considered there are identical restructured paraphrases of (6) and (7) which therefore have the same,

[15] See Fred Dretske, 'Contrastive Statements', *Philosophical Review* LXXXI (1972), pp. 411–37.

truth-value. But this violates assumption I since, according to the latter, if (6) and (7) have different truth-values so must their restructured paraphrases.

6   THE QUESTION VIEW

The ordered pair view may have advantages over other object views, but it is not acceptable because assumption I is not satisfied. Different questions about an event can change the truth-value of the claim that an explainer explains the event under a given description. Perhaps then an ordered triple consisting of an event (state of affairs, etc.), a description, and a question can be an object of explanation.

What conditions would the third member of this ordered triple need to satisfy? The question must be about the event under the description d. We might say that a question Q presupposes a description d if the soundness of the question, i.e. its having an answer that is true, semantically requires the truth of the description d. 'Why did Kissinger warmly shake hands with Le Duc Tho?' and 'How did Kissinger warmly shake hands with Le Duc Tho?' both presuppose the description 'Kissinger warmly shook hands with Le Duc Tho'. One requirement, then, is that the question in the ordered triple must presuppose the description d, the second member of that triple. Secondly, since questions can differ with different contrasts, contrasts, if any, that are being made when a question is raised are to be explicitly indicated. On the present view, then, an ordered triple (e, d, Q) is an object of explanation iff

1   e is an event (state of affairs, etc.)
2   d is a description of e
3   Q is a question that presupposes d
4   contrasts, if any, are explicit in Q.

Two ordered triples (x, y, z) and (a, b, c) are identical iff $x = a, y = b$, and $z = c$. According to the present view, if *events* (states of affairs, etc.) in two ordered triples are different what is explained is different. To explain why Kissinger shook hands yesterday is not to explain the same thing as why he shook hands today. Similarly if the *questions* are different the objects of explanation are different. To explain why Kissinger shook hands and to explain how Kissinger shook hands is not to explain the same thing. And if *descriptions* are different the objects are too.

However, if we are to characterize a thing-transparent use of 'explains' a problem now emerges which will force us to take a closer look at the nature of descriptions and questions. Let

event $e =$ Jones' getting indigestion
description $d_1 =$ Jones got indigestion
description $d_2 =$ the best biographer of Freud got indigestion
question $Q_1 =$ why did Jones get indigestion?
question $Q_2 =$ why did the best biographer of Freud get indigestion?

Assume that Jones is the best biographer of Freud. In accordance with a thing-transparent use of 'explains' to explain why Jones got indigestion and to explain why the best biographer of Freud got indigestion is to explain the same thing. If we construe this as the claim that the same objects of explanation are involved and that these are ordered triples of the sort we have described, then we must suppose that $(e, d_1, Q_1) = (e, d_2, Q_2)$. But this cannot be if descriptions and questions are sentences. The sentence 'Jones got indigestion' is not identical with the sentence 'the best biographer of Freud got indigestion'. Nor is the sentence 'why did Jones get indigestion?' identical with the sentence 'why did the best biographer of Freud get indigestion?' If objects of explanation (e, d_1, Q_1) and (e, d_2, Q_2) are identical then what is needed is a different interpretation of descriptions and questions.

Before searching for such an interpretation let us contrast this case with the following description of, and question about, the event e above:

description d_3: the person who took pills to prevent
indigestion got indigestion
question Q_3: why did the person who took pills to prevent
indigestion get indigestion?

And let us suppose that Jones is the person who took pills to
prevent indigestion. To explain why Jones got indigestion, we
might say, is not necessarily to explain why the person who took
pills to prevent indigestion got indigestion. In the latter case
what may be wanted is an explanation of why the person got
indigestion despite his having taken measures to prevent it,
while in the former case this is not what is wanted. We might
say then that a sentence such as

Smith explained why the person who took pills to prevent
indigestion got indigestion

has two readings. On one, the thing-transparent reading, the
sentence remains true when any other expression that refers
to Jones is substituted for 'the person who took pills to prevent
indigestion'. On the other, the thing-opaque reading, the sentence does not in general remain true under such substitutions.
On the latter reading the particular expression used in describing the thing makes a difference for what is explained and
hence for the truth-value of the explanatory sentence. In what
follows the concepts of description and question will be re-
interpreted so as to be able to handle both of these readings of
explanatory sentences.

Let us say that a description is not a sentence but what a
speaker provides when, using a sentence, he describes some
event (state of affairs, etc.), and let us try to specify a criterion
for deciding when descriptions provided by speakers using
two sentences are the same. I shall do so only for sentences of a
fairly simple kind, ones in which a predicate is ascribed to a
particular (or a relational predicate to an n-tuple of particulars).
Such sentences, which include 'Jones got indigestion', and
'the best biographer of Freud got indigestion on Tuesday in
London', will be called *basic descriptive* sentences. In order to

accommodate both thing-transparent and thing-opaque readings of explanatory sentences two kinds of descriptions will need to be distinguished. In using two basic descriptive sentences in describing the same event (state of affairs, etc.) I shall say that two speakers provide the same *A*-description iff in using the sentences they are ascribing the same predicate to the same particular(s); and they provide the same *B*-description iff in addition the particular is designated by the same expression in both sentences. In using the sentences 'Jones got indigestion' and 'the best biographer of Freud got indigestion' in describing the same event two speakers provide the same *A*-description since they ascribe the predicate 'got indigestion' to the same person, but they do not provide the same *B*-description since that person is designated by different expressions in these sentences.[16]

A similar procedure will be followed with questions. A question will be treated not as an interrogative sentence but as what is expressed by a speaker using an interrogative sentence when that speaker is asking about some event (state of affairs, etc.). Let us try to specify a criterion for deciding when questions

[16] I shall leave open the question of whether logically or semantically equivalent predicates can be substituted and the description remain the same, and also the question of whether speakers using logically or semantically equivalent sentences in describing the same event provide the same description. What is definitely precluded are predicates that are simply coextensive or predicates which contingently express the same property, like 'blue' and 'the colour of the sky'. Also in using a sentence such as 'the best biographer of Freud got indigestion' in providing a description a speaker must be using the definite description as a referring expression, but he need not know that the person to whom he has referred is Jones. That is, in using two sentences in describing the same event speakers may provide the same description even though they do not know that they have. Kim has a proof from which it follows that if the substitution of co-referring terms does not change the description and if logically equivalent sentences used in describing the same event provide the same description, then in using any two true sentences to describe the same event speakers will be providing the same description of that event. But, as Kim recognizes, this proof depends on treating all definite descriptions as referring, something he himself ends up by disallowing. See Jaegwon Kim, 'Events and their Descriptions: Some Considerations', *Essays in Honour of Carl G. Hempel*, pp. 198–215.

expressed by speakers using two interrogative sentences are the same. As in the case of descriptions this will be done only for interrogative sentences of a fairly simple kind, ones that will be called *basic interrogative* sentences. They begin with an interrogative word or phrase ('why', 'for what reason') followed by a phrase that can be transformed into a basic descriptive sentence, for example 'Why did Jones get indigestion?' (This sentence will be said to *contain* the descriptive sentence 'Jones got indigestion'.) The identity criterion is this. In using two basic interrogative sentences S_1 and S_2 in asking about events (states of affairs, etc.) speakers express the same question iff

1   S_1 and S_2 contain basic descriptive sentences which the speakers would use in providing the same description

2   any correct answer to one question is a correct answer to the other.

In (1) if 'description' is understood as 'A-description' then we can speak of the question involved as an A-question. If 'description' is understood as 'B-description' then the question is a B-question. For example according to the above conditions, in using the basic interrogative sentences 'why did Jones get indigestion?' and 'why did the best biographer of Freud get indigestion?' in asking about the same event speakers would be expressing the same A-question but not the same B-question. In what follows I shall confine attention to the thing-transparent use of 'explains' and for this purpose descriptions and questions will always be understood as A-descriptions and A-questions.

Let us now return to our previous example. Using the bracket notation $[\ ]_e$ for the description of e provided by a speaker using the sentence inside the brackets, and the parenthesis notation $(\ )_e$ for the question about e expressed by a speaker using the sentence inside the parentheses, we can recast this example as follows

$e =$ Jones' getting indigestion
$d_1 =$ ['Jones got indigestion']$_e$
$d_2 =$ ['the best biographer of Freud got indigestion']$_e$

Q_1 = ('why did Jones get indigestion ?')$_e$
Q_2 = ('why did the best biographer of Freud get indigestion ?')$_e$

Here the objects of explanation (e, d_1, Q_1) and (e, d_2, Q_2) are the same since $d_1 = d_2$ and $Q_1 = Q_2$.

At this point another problem appears. Let

e = Jones' getting indigestion
d_1 = ['Jones *got indigestion* on Tuesday at home']$_e$
d_2 = ['Jones *got indigestion* in January in London']$_e$
Q_1 = (why did Jones *get indigestion* on Tuesday at home ?')$_e$
Q_2 = ('why did Jones *get indigestion* in January in London ?')$_e$

To explain why Jones *got indigestion* on Tuesday at home and to explain why Jones *got indigestion* in January in London, is, it seems plausible to say, to explain the same thing, if Jones got indigestion on Tuesday, at home, in January, in London. (We are not here emphasizing 'on Tuesday' or 'at home', etc.; otherwise what is explained would be different.) Yet given our previous criterion descriptions d_1 and d_2 are different since different predicates are being ascribed to Jones. So the objects of explanation are different.

To avoid this conclusion our proposals will need to be modified. Let us say that in using a basic descriptive sentence S a speaker provides an *emphasis-description* (E-description) iff he is using S to describe some event (state of affairs, etc.) and some component of S is emphasized. We can then say that in using basic descriptive sentences S_1 and S_2 speakers provide the same E-description iff

1   in using S_1 and S_2 they describe the same event, etc.
2   in using S_1 and S_2 they are ascribing predicates to the same item(s).[17]

[17] We are here characterizing E-descriptions that are also A-descriptions. For B-descriptions condition (2) would need to be modified by adding:
'which is designated by the same expression in S_1 and S_2'.

3   any component in one sentence that is emphasized appears in the other and is emphasized. For example 'Jones *got indigestion* on Tuesday at home' and 'Jones *got indigestion* in January in London'.

4   the same basic verbial component appears in both sentences. By the basic verbial component I mean that component of the verb phrase with adverbs, prepositional phrases, and modifying clauses deleted that by itself can be ascribed to the subject. For example in 'Jones got indigestion on Tuesday at home after eating dinner' it is 'got indigestion'.[18]

In using a basic descriptive sentence S a speaker provides a *description* iff he is using S to describe some event (state of affairs, etc.), and we can modify our earlier proposal and say that in using two basic descriptive sentences S_1 and S_2 speakers provide the same description iff conditions (1), (2) and (4) above are satisfied.

In accordance with these conditions, in using the following sentences to describe the same event speakers would provide the same E-description: 'Jones *got indigestion*', 'the best biographer of Freud *got indigestion*', 'Jones *got indigestion* on Tuesday at home'. However, these sentences could not be used to provide the same E-descriptions as any of the following: 'Smith *got indigestion*' (where it is assumed that Smith ≠ Jones, so that conditions (1) and (2) are violated), 'Jones got indigestion *on Tuesday*' (condition (3) is violated), 'Jones *ate parsnips* on Tuesday' (conditions (1), (3), and (4) are violated). Note that 'Jones *got indigestion* on Tuesday at home' and 'Jones got indigestion *on Tuesday*' could be used to provide the same description but not the same E-description, since conditions (1), (2), and (4) but not (3) are satisfied. Obviously the term 'description' is now being used in a very special sense. We might

---

[18] This condition precludes the following sentences from being used to provide the same E-description: 'this gas has a temperature of $300°K$' and 'this gas has a mean molecular kinetic energy of $6·21 \times 10^{-21}$ joule'. Even though conditions (1) to (3) are satisfied condition (4) is not since the basic verbial components are different.

say that *for explanatory purposes* these two sentences can be used in providing the same description.

The account of questions and the two conditions for expressing the same question given on p. 22 can be retained. But we can now add that in using an interrogative sentence S a speaker expresses an E-question about an event if he expresses a question about that event and if in S one or more words (other than the interrogative words) are emphasized. And in using two basic interrogative sentences S_1 and S_2 in asking about events speakers express the same E-question iff condition (2) for expressing the same question is satisfied, and condition (1) is satisfied with 'same E-description' replacing 'same description.'

According to these proposals, in using the interrogative sentences 'why did Jones *get indigestion* on Tuesday at home ?' and 'why did Jones *get indigestion* in January in London ?' in asking about the same event speakers would express the same question and also the same E-question. But they would not express the same question or E-question using these as they would using 'why did Jones *eat parsnips* on Tuesday ?' (descriptive sentence contained would not be used to provide the same description or E-description), or 'why did Jones *get indigestion* from eating parsnips ?' (not every correct answer to the former is a correct answer to the latter; for example 'Jones got indigestion from eating parsnips' is not a correct answer to the latter though it might be to the former).

A word is in order about condition (2) on p. 22 for expressing the same (E-) question. Let us speak of sentences and phrases as being answers to a question and say that two sentences or phrases are *different* answers to a question if (but not only if) a component appears in one but not in the other or if a component appears emphasized in one but not in the other. For example, 'because Jones *ate parsnips* on Tuesday', 'because Jones *ate parsnips*', and 'because Jones ate parsnips *on Tuesday*' are different answers to the question that would be expressed by asking 'why did Jones *get indigestion* on Tuesday ?' Two

BE

different answers to a question may both be correct. In the present case, let us suppose, 'because Jones *ate parsnips* on Tuesday' and 'because Jones *ate parsnips*' are both correct answers to the question. However, two sentences containing the same components but different emphases may be such that one is a correct answer to a question and the other incorrect. 'Because Jones *ate parsnips* on Tuesday' is a correct answer and 'because Jones ate parsnips *on Tuesday*' is (in the present case) an incorrect answer to 'why did Jones *get indigestion* on Tuesday?' since it incorrectly implies that it was the day on which he ate them that caused his indigestion.

If in using two basic interrogative sentences speakers express the same E-question then they express the same question (since if basic descriptive sentences are used to provide the same E-description then they are used to provide the same description). However, the converse does not hold in general. Consider the following basic interrogative sentences:

1   where did Jones *get indigestion* on Tuesday?
2   where did Jones get indigestion *on Tuesday*?

In using (1) and (2) in asking about the same event speakers would not express the same E-question since in using the descriptive sentences (1) and (2) contain they would not provide the same E-description. Nevertheless, in using (1) and (2) in asking about the same event speakers would express the same question since the basic descriptive sentences they contain would be used to provide the same description, and any correct answer to one is a correct answer to the other (for example 'at home, in London'). No doubt (1) and (2) would be uttered in different circumstances by persons with different pieces of knowledge. (1) might be uttered by someone who knows where Jones got a headache on Tuesday but not where he got indigestion ('I know where he got a headache on Tuesday, but where did he *get indigestion* on Tuesday?'). (2) might be uttered by someone who knows where Jones got indigestion on Monday but not on Tuesday.

Compare interrogative sentences (1) and (2) with the following:

3   why did Jones *get indigestion* on Tuesday?
4   why did Jones get indigestion *on Tuesday*?

In using (3) and (4) in asking about the same event, as in the case of using (1) and (2), speakers would not express the same *E-question*, but unlike the case of (1) and (2), in using (3) and (4) they would not express the same question. The reason is that we are assuming that 'because he ate parsnips *on Tuesday*' is a correct answer to the question a speaker would express using (4) but not a correct answer to the question a speaker would express using (3).

Let us say that in using interrogative sentences such as (3) and (4) a speaker expresses a *type I* question, and that in using interrogative sentences such as (1) and (2) he expresses a *type II* question. A question expressed using an interrogative sentence is a type I question iff a change in emphasis in the interrogative sentence would change the question expressed. A question expressed using an interrogative sentence is a type II question iff a change in emphasis would not change the question expressed. A type I or II question expressed using an interrogative sentence may also be an *E*-question. (3) and (4) would be used in expressing type I *E*-questions, (1) and (2) in expressing type II *E*-questions.

Let us say that an (*E*-) question *Q presupposes* an (*E*-) description *d* iff there is an interrogative sentence which a speaker could use in expressing *Q* and this interrogative sentence contains a descriptive sentence that speakers would use in providing an (*E*-) description *d*. In general, it follows from condition (1) on p. 22 that any basic interrogative sentence a speaker uses in expressing the same (*E*-) question contains descriptive sentences he would use in providing the same (*E*-) description. So we can also say that an (*E*-) question *Q* presupposes an (*E*-) description *d* iff any basic interrogative sentence a speaker uses in expressing *Q* contains a basic descriptive sentence he would use in providing an (*E-*) description *d*.

On the present view an ordered triple that is an object of explanation consists of an event (state of affairs, etc.), a description of that event, and a question which presupposes that description. But as we now construe the latter items there will be redundancy in the ordered triple, since if we identify a question we identify a description and an event. Accordingly, instead of saying that an ordered triple is an object of explanation we can simplify the view by identifying the object with the third member of that triple, the question. Let us write this question as $Q(e, d)$ which, we can say, is an object of explanation if

1   e is an event (state of affairs, etc.)
2   d is a description of e
3    $Q(e, d)$ is a question about e that presupposes d
4   If $Q(e, d)$ is a type I question then it is also an *E*-question and d is an *E*-description; if $Q(e, d)$ is a type II question then it may or may not be an *E*-question and d an *E*-description.[19]

It is now possible to formulate

The Question View: Any preferred explanatory sentence is paraphrasable into a restructured one whose object term refers to a question $Q(e, d)$ satisfying conditions (1) to (4) above. (And any preferred explanatory sentence is restructured only if its object term refers to such a question.)

A question can have both direct and indirect formulations. Or to put it in terms we have been using, a speaker may express the same question using both direct and indirect interrogatives. 'Why did Socrates die?' and 'why Socrates died' can be used to express the same question. The only difference is that the direct interrogative can be used by itself to express a question,

---

[19] These conditions are meant to hold whether or not the descriptive sentence used to provide description *d* and the interrogative sentence used to express question Q are basic, although criteria for providing the same (*E-*) description and for expressing the same (*E-*) question have been proposed only for basic sentences.

whereas the indirect interrogative must be embedded in a sentence. On object-of-explanation views restructured occurrences of 'explains' are followed by expressions purporting to refer to an object of explanation. A defender of the question view might claim that an indirect interrogative like 'why Socrates died' purports to refer to a question. If so, the explanatory sentence 'Plato explained why Socrates died' is restructured. Or he might claim that the indirect interrogative 'why Socrates died' does not purport to refer at all, in which case the explanatory sentence 'Plato explained why Socrates died' is not restructured. On the latter view, to obtain a paraphrase of this sentence that is restructured we can write 'Plato explained the question of (or, as to) why Socrates died', in which the expression following 'explained' does purport to refer to a question.

7 IS THE QUESTION VIEW ADEQUATE?

In section 1 an object of explanation was characterized as something that can be said to be explained, as something a knowledge of which puts one in a position to seek an explanans, and as something concerning which general theories of explanation can be developed. Do questions satisfy these criteria? Consider the explanatory sentence

Plato explained why Socrates died.

A question theorist may say that this is restructured, in which case what Plato explained is why Socrates died, i.e. a question. Or he may say that it is not restructured, but that 'Plato explained the question of (or, as to) why Socrates died' is, in which case, again, it is a question that is explained. (Or, as will be indicated in section 9, he may propose a different restructuring for explanatory sentences in which questions are objects of explanation.) Second, if we know what question is being raised then we know what event is being referred to, under what description, and what it is about that event under that description that is being questioned. On the present view, if we know these things we are in a position to seek an explanans. At least the question theorist is justified in saying that on this score his viewpoint is superior to those of the other object views mentioned. If we know what question it is that is being raised about a given event under a given description we are in a better position to know what sort of explanans to seek than if we know only the event, or the description, or even both. Finally, the issue of whether questions are objects with respect to which general theories of explanation can be developed would take us far afield of the present topic. Suffice it to note that attempts have been made to develop general theories of explanation that supply truth conditions for explanatory sentences taking questions as objects, where such sentences are treated as more basic than other explanatory sentences.[20]

In defence of the question view we can also show that the kind of example that earlier proved fatal to other object views does not do so in the present case. Consider the sentences

1   Smith explained why *Kissinger* warmly shook hands with Le Duc Tho

2   Smith explained why Kissinger warmly shook hands with *Le Duc Tho*

3   Smith explained where Kissinger warmly shook hands with Le Duc Tho,

and let us suppose that (1) is true but (2) and (3) are false. On each of the other object views we have considered there are restructured paraphrases of (1) to (3) which are identical and hence have the same truth-value. But this is precluded by assumption I which requires that if (1) to (3) do not all have the same truth-value then neither can their restructured paraphrases. On the question view, however, (1) to (3) are all restructured and since the questions are different in each case

[20] See Sylvain Bromberger, 'An Approach to Explanation', *Analytical Philosophy*, ed. R. J. Butler (Oxford, 1965), II, 72–105, and my *Law and Explanation*, ch. 4.

this view does not require these sentences to share a common truth-value. (Even if a defender of the question view does not consider (1) to (3) to be restructured, but requires 'Smith explained the question of . . .', still the questions referred to in such restructured paraphrases would all be different; and again (1) to (3) would not be required to have the same truth-value.)

Assumption II of object-of-explanation views is that in restructured explanatory sentences terms purporting to refer to objects of explanation and explainers occur purely referentially. One sub-condition is that from a restructured explanatory sentence we may infer sentences of the form '$(\exists x)(\ldots$ explains $x)$' and '$(\exists x)(x$ explains $\ldots)$'—a condition satisfied by the question view. Another is that if one restructured explanatory sentence can be obtained from another by substituting co-referring terms for the explainer term or the object term then the explanatory sentences have the same truth-value. In section 9 an example will be proposed that casts at least some doubt on whether the question view (without some modification) satisfies this condition. But we can show that the kind of example that was used to embarrass the non-linguistic view of section 2 does not do so in the present case. Consider

4   Plato explained why Socrates *died*
5   Plato explained why Socrates died *from drinking hemlock*,

and let us suppose that (4) is true but (5) is false. On the non-linguistic view (4) and (5) have the restructured paraphrases

6   Plato explained Socrates' death
7   Plato explained Socrates' death from drinking hemlock.

Now assuming that (6) is true but (7) is false (i.e. assuming that (6) and (7) are adequate paraphrases of (4) and (5)), and assuming that 'Socrates' death' and 'Socrates' death from drinking hemlock' are co-referring expressions, assumption IIB is violated. This is because (6) can be obtained from (7) and conversely by substitution of co-referring expressions for the object of explanation, and this requires (6) and (7) to have the same truth-value, which by hypothesis they do not. The question view avoids this consequence since it does not take (6) and (7) to be restructured paraphrases of (4) and (5).

It might be asked whether stronger versions of the question view are plausible. For one thing, can the parenthesized material in the formulation of the question view at the end of section 6, i.e.

8   Any preferred explanatory sentence is restructured only if its object term refers to a question

be defended? Such a position is, I think, plausible, especially in view of the fact that other objects we have considered—non-linguistic items such as events, etc., descriptions, and ordered pairs of these—cannot be taken to be objects of explanation in preferred explanatory sentences such as (1) to (3) above, on pain of violating assumption I. Unless other types of objects are proposed for such sentences that can reasonably be called objects of explanation and that don't lead to violations of assumption I, (8) seems plausible indeed.

Might an even stronger version be defended? Could it be claimed that

9   Any explanatory sentence whatever (preferred or not) is paraphrasable into a restructured one whose object term refers to a question. And any explanatory sentence is restructured only if its object term refers to a question

or at least that

10   Any explanatory sentence whatever is paraphrasable into a restructured one whose object term refers to a question?

The latter has a fighting chance. To defend it one would need to show how to handle such examples as

11   Albert explained the first sentence of the *Declaration of Independence*

12   George explained the company policy
13   Robert explained that he forgot his umbrella.

In (11) explaining this sentence would normally be understood as explaining what it means. In this case *e* could be taken to be the state of affairs of sentence *S*'s having a meaning, *d* to be a description of this state of affairs provided by a speaker using the sentence '*S* has a meaning', and *Q* (*e*, *d*), an object of explanation for (11), to be the question ('what meaning does *S* have?')$_e$. In (12) *e* could be construed as the state of affairs of there being a company policy, *d* as a description of this state of affairs, and *Q* (*e*, *d*) as the question ('what is the company policy?')$_e$. 'Explained that . . .' is different from other indirect forms since what replaces the blank is not an expression designating an object of explanation but rather (an expression designating) the explanans. If Robert explained that he forgot his umbrella it was not the forgetting of his umbrella that he explained but something else (for example his getting wet), and this he explained by citing the fact that he forgot his umbrella. On the question view, then, (13) can be paraphrased as '($\exists Q$) (Robert explained *Q*. Robert's explanans for *Q* was that he forgot his umbrella.)'

Obviously this is not enough to establish (10), but it does indicate at least how one might proceed to defend it. Whether the even stronger (9) can be defended I shall not venture to say here. In what follows I want to consider the consequence for one traditional theory of explanation if the question view with (8) included (or the much stronger (9)) is adopted. So when I speak of the question view I shall be referring to these versions.

8 IMPLICATIONS OF THE QUESTION VIEW FOR THE D–N MODEL OF EXPLANATION

Theories of explanation concentrate much more on the explanans than on what is explained. But attention to the latter will, I think, suggest some interesting things about the former.

Here, however, I shall only mention one theory of the explanans, the Deductive–Nomological model, and note some implications for this theory if explanatory sentences are construed in accordance with the question view.

On the D–N model the explanatory sentence '*A* explains why Socrates died' is true if *A* is a set of sentences containing laws and initial condition sentences and *A* entails some sentence, say 'Socrates died', which describes the event of Socrates' death.[21] But on the question view, this is not sufficient for the truth of '*A* explains why Socrates died'. That is, the following cannot be accepted:

1   '*A* explains Q (*e, d*)' is true if *A* contains laws and initial condition sentences and *A* entails *D*, where *D* is a descriptive sentence used by a speaker in providing description *d* of event *e*.

Let

$$e = \text{the event of Socrates' death}$$
$$d = [\text{'Socrates died'}]_e$$
$$Q_1 (e, d) = (\text{'why did Socrates die ?'})_e$$
$$Q_2 (e, d) = (\text{'how did Socrates die ?'})_e$$
$$Q_3 (e, d) = (\text{'where did Socrates die ?'})_e$$

Suppose we explain Q_1 (*e, d*), i.e. why Socrates died, by appeal to an explanans *A*, which, to satisfy D–N theorists, contains laws and initial condition sentences that entail the sentence 'Socrates died'. But if *A* explains Q_1 (*e, d*) simply in virtue of the fact that it contains laws and initial condition sentences that entail a sentence a speaker would use in providing description *d* of event *e*, i.e. if (1) is true, then *A* must also explain Q_2 (*e, d*) and Q_3 (*e, d*), since the descriptions of *e* contained in each of these questions is the same. But clearly while *A* may explain Q_1 (*e, d*) it might not explain Q_2 (*e, d*) or Q_3 (*e, d*). Although *A*

---

[21] *A* is here the explanans. It may or may not also be taken to be the explainer. If it is not—if, for example, explainers are always persons— then instead of '*A* explains why Socrates died' we could use 'By citing (explanans) *A* person *P* explains why Socrates died'.

might explain why Socrates died it might not explain how he died or where he died. To explain $Q(e, d)$ it is not sufficient to deduce a sentence a speaker uses in providing description d from laws and initial condition sentences. The reason is, of course, that different questions about e are possible, all of which presuppose the same description.

Thinking of the object of explanation as a question will also help us to see why some standard counterexamples to the D–N model are counterexamples. Consider the following two cases.

1 We see a fire engine that is black and demand an explanation. Here is one that satisfies the D–N model:

> This fire engine is the same colour as that crow
> All crows are black
> _____
> This fire engine is black

2 A bridge collapses and we demand an explanation. Here is one that satisfies the D–N model:

> An engineer with years of training and experience examined the bridge and said that it would collapse.
> Whenever an engineer with years of training and experience examines a bridge and says that it will collapse then it does.
> _____
> The bridge collapsed.

These are unacceptable as explanations, and the reason is that they don't answer the questions most likely to be asked when explanations are demanded in these cases. In case 1 the most likely question to be asked by someone who sees a black fire engine and demands an explanation is 'for what reason is this fire engine black?' And that question is not answered by the D–N explanans. In case 2 the likely question is 'what caused the bridge to collapse?'—a question to which the D–N explanans fails to address itself.

Finally, the question view has implications for the thesis, supported by D–N theorists, that there is a symmetry between explanation and prediction. Suppose we explain why Jones got indigestion on Tuesday in London at home after eating dinner by saying that he has ulcers and ate parsnips; and to satisfy the D–N craving for laws let us suppose that it is a law that all those with ulcers who eat parsnips get indigestion, and we cite this law in our explanation. Still the D–N theorist won't be satisfied because the explanation-prediction symmetry is not present. If we had known in advance that Jones who has ulcers would eat parsnips we could have predicted that he would get indigestion, but we could not have predicted the event in question. We could not have predicted that he would do so on Tuesday, in London, at home, after eating dinner. By D–N standards, then, we have not explained this event.

The question view disallows this conclusion. Let

$e =$ the event of Jones' getting indigestion on Tuesday in London at home after eating dinner

$d_1 =$ ['Jones *got indigestion* on Tuesday in London at home after eating dinner']$_e$

$d_2 =$ ['Jones got indigestion *on Tuesday* in London, etc.']$_e$

$d_3 =$ ['Jones got indigestion on Tuesday *in London*, etc.']$_e$

$Q_1 (e, d_1) =$ ('why did Jones *get indigestion*, etc.')$_e$

$Q_2 (e, d_2) =$ ('why did Jones get indigestion *on Tuesday*, etc. ?')$_e$

$Q_3 (e, d_3) =$ ('why did Jones get indigestion on Tuesday *in London*, etc ?')$_e$

The explanans 'because Jones has ulcers and ate parsnips and all people under these conditions get indigestion' might be a reasonably good explanation of $Q_1 (e, d_1)$ but not of $Q_2 (e, d_2)$ or of $Q_3 (e, d_3)$. And since $Q_1 \neq Q_2 \neq Q_3$ this is perfectly legitimate. Our explanation may explain why Jones *got indigestion* on Tuesday but not why he got indigestion *on Tuesday*. If so then the symmetry claim must be rejected. The explanans above explains $Q_1 (e, d_1)$ but is not such that if in advance we had known the facts described by this explanans would have allowed us to predict the occurrence of e as described by d_1. If we had known that Jones who has ulcers would eat parsnips we could not have predicted that he would get indigestion on Tuesday in London.

This means that the converse of (1), as well as (1), must be rejected, i.e.

2   'A explains Q (e, d)' is true *only if* A contains laws and initial condition sentences and A entails D, where D is a descriptive sentence used by a speaker in providing description d of event e.

If 'because Jones has ulcers and ate parsnips and all people under these conditions get indigestion' is a reasonable explanation of Q_1 (e, d_1), then (2) above is false. This is because this explanans, though it contains a law and an initial condition statement, does not entail the descriptive sentence 'Jones *got indigestion* on Tuesday in London at home after eating dinner' used by a speaker in providing description d_1 of event e.

9   A NO-OBJECT VIEW

All of the views elaborated so far suppose that an explanatory sentence expresses a relationship between an explainer and some object of explanation. But objections can be raised against such views. Ontological purists, for example, will object that the question view commits users of explanatory sentences to the existence of such abstract objects as questions and descriptions. And there is the following difficulty that the question view (as well as the sentence view) faces.[22] Consider once again the explanatory sentence

1   Plato explained why Socrates died.

Assume that this is restructured and that, in accordance with assumption IIB, the object position is referentially transparent

[22] Here I am indebted to George Wilson.

and can be filled by an expression designating the same question. Then if (1) is true so is

2   Plato explained the question 'why did Socrates die ?'

But, it might be argued (1) and (2) have totally different meanings and indeed different truth-values, since (2) is equivalent to

3   Plato explained the meaning of the question 'why did Socrates die ?'

and (3), we may suppose, is false even though (1) is true. A similar problem will arise if we say that (1) is unrestructured and produce a restructured paraphrase such as 'Plato explained the question of why Socrates died'.

There are, I think, three responses someone sympathetic to the question view might make to the latter objection. First, he might deny that (2) must be construed as being equivalent to (3). He might agree that there is a use of (2) that makes (2) and (3) equivalent, but he might urge that (2) can also be understood as being equivalent to (1) but not to (3). (Such a defence would seem to me to be quite weak.) Alternatively, he might modify the concept of a *restructured* occurrence of 'explains' as one that occurs in the context '. . . explains by answering the question . . .'. Here 'explains by answering', rather than simply 'explains', is understood as relating an explainer and something else. When this expression is followed by an expression purporting to refer to a question then the occurrence of 'explains' is restructured. With this proposal assumption IIA would need to be modified by replacing 'explains' by 'explains by answering', but no changes in the formulations of assumptions I or IIB would be required. Finally, someone sympathetic to the question view might abandon the idea that explanatory sentences relate an explainer and an object of explanation while still holding that such sentences are paraphrasable into sentences of the form '. . . explains . . .' in which the second blank is filled by an indirect interrogative. It is this latter possibility that will now be examined.

Scheffler considers (but does not adopt) the view that the term 'explains' functions as an operator on sentences to produce predicates.[23] Although he thinks that such predicates apply to sentences (i.e. that the explainer is a sentence) we need not be committed to this position. In the sentence

1   Plato explained why Socrates died

we can treat 'explained why' as a sentential operator which operates on the sentence 'Socrates died' to yield the predicate 'explained why Socrates died', and the latter predicate is being ascribed to Plato by (1).

Scheffler goes on to consider two versions of this view. According to the first, the predicate 'explained why Socrates died' is to be treated as dyadic, so that (1) has the form

4    $aEFb$

in which 'a' is to be replaced by 'Plato' and 'b' by 'Socrates' in order to obtain (1). According to this version, from a sentence of form (4) we can infer one of form

5    $(\exists x)\,(\exists y)\,(xEFy)$,

so that from (1) we can infer 'someone explained why someone died'.

According to the second version, the predicate 'explained why Socrates died' is to be treated as monadic, so that (1) has the form

6   aEp

in which 'Ep' is to be replaced by 'explained why Socrates died' in order to obtain (1). And from a sentence of the form (6) we cannot infer one of the form (5).

Of these two versions Scheffler prefers the second, since in accordance with the first the b-position in (4) is referentially

[23] Scheffler, op. cit., pp. 62ff.

transparent, and Scheffler thinks that this leads to difficulties. I believe that he is mistaken on this point, since, as urged above, there is a thing-transparent reading of explanatory sentences. Accordingly, I shall assume that the first version of the no-object view is a plausible no-object view.

Let us try to develop such a view further. The problem with it is similar to problems that led to the formulation of the question view. By analogy with assumption IIB it might seem plausible to say

III   If two sentences S_1 and S_2 contain an 'explains' operator which operates on the same sentence to produce the same predicate, and S_1 and S_2 ascribe this predicate to the same individual, then S_1 and S_2 have the same truth-value.

For example, in accordance with III the following sentences must have the same truth-value:

Plato explained why Socrates died

The author of the *Republic* explained why Socrates died.

But now consider the following sentences

7   Smith explained why Jones got indigestion
8   Smith explained where Jones got indigestion,

and let us suppose that (7) is true but (8) is false. If (7) and (8) are construed as being formed by letting 'explain' operate on the sentence 'Jones got indigestion' to form a predicate, then the same predicate is formed in each, since the sentence 'Jones got indigestion' is the same in each, and this predicate is ascribed to the same individual. Therefore, by III, (7) and (8) must have the same truth-value, which, by hypothesis, they do not. The problem, of course, is that this view fails to take account of the different questions in (7) and (8). We could build the questions into the explains-operators, thus yielding 'explains why', 'explains where', etc., as different operators, though a problem would still remain analogous to the one earlier because of the possibility of different emphases. So let me consider another alternative more in line with previous strategy (though not an object-of-explanation view).

In dealing with object-of-explanation views we spoke of restructured occurrences of 'explains' as those followed by an expression purporting to refer to an object of explanation. On the present no-object view let us speak of a restructured occurrence of 'explains' as one followed by an *indirect interrogative*. With this change we can retain assumption I. That is, we can say that when a speaker uses an explanatory sentence S_1 there is some explanatory sentence S_2 with a restructured occurrence of 'explains' that expresses what the speaker means to be saying and has the same truth-value as S_1. And we can speak of a restructured explanatory sentence as one in which the occurrence of 'explains' is restructured.

In restructured explanatory sentences, on this view, 'explains' is an operator on *interrogative sentences* which yields a predicate applicable to the explainer. The interrogative sentences may contain emphases. Thus consider the interrogative sentence

9   why did Jones *get indigestion* on Tuesday?

A predicate is formed by letting 'explains' operate on (9). We might suppose that 'explains' operates in such a way as to transform (9) into an *indirect* interrogative, so that we obtain the predicate

10   explains why Jones *got indigestion* on Tuesday.

Now consider the interrogative sentence

11   why did Jones *get indigestion* in January?

and let us suppose that

12   the event of Jones' getting indigestion on Tuesday = the event of Jones' getting indigestion in January.

On the question view developed earlier, from

13   *A* explained why Jones *got indigestion* on Tuesday

together with (12) we can infer

14   *A* explained why Jones *got indigestion* in January

since, we may suppose, 'why did Jones *get indigestion* on Tuesday?' and 'why did Jones *get indigestion* in January?' are being used in expressing the same question. However, on the present no-object view such an inference cannot be made without the introduction of additional rules.

What sort of rules will these be? According to the present view in a restructured explanatory sentence 'explains . . .' is a predicate formed by treating 'explains' as an operator on an interrogative sentence q (small q will now be used as a variable ranging over interrogative sentences). In order to obtain a sentence like (14) from (12) and (13) we need a rule such as this:

> *Rule R:* If the interrogative sentences q_1 and q_2 are used by a speaker to express the same type I E-question (see section 6), then 'A explains q_1' is true iff 'A explains q_2' is true (where 'explains q' is a predicate formed by letting the explanation operator operate on the interrogative sentence q).

For example, given (12) we can say that (9) and (11) would be used to express the same type I E-question. Applying the rule above we are permitted to infer (14) from (13), and conversely. By analogy with assumption IIB we have the following assumption, from which rule R follows:

> IV If two sentences S_1 and S_2 contain an 'explains' operator which operates on interrogative sentences q_1 and q_2 to produce predicates P_1 and P_2, and if q_1 and q_2 are used to express the same type I E-question, and if S_1 is obtainable from S_2 by substituting P_1 for P_2 or by substituting a co-referring term for the explainer term, then S_1 and S_2 have the same truth-value.

In accordance with IV, (13) and (14) have the same truth-value.

On the present no-object view explanatory sentences such as (1), (13), and (14) do not entail the existence of objects of explanation, for example the existence of questions. Nor does the use of rule R in making inferences (from, say, (13) to (14)) commit one to the existence of questions (or events, or descriptions). To use R in generating (15) from (13) we must suppose that (9) and (11) are being used to express the same (type I *E*-) question, which in turn involves the assumption that 'Jones *got indigestion* on Tuesday' and 'Jones *got indigestion* in January' are being used in providing the same (*E*-) description, which, finally, involves the assumption that the latter sentences are being used in describing the same event. But just as one can say that two sentences have the same meaning without supposing that there exists an object called 'a meaning' which they both have, so a defender of a no-object view can speak of two interrogative sentences as being used to express the same question without supposing that there exists an object called 'a question' which they are both used to express. And just as the defender of the no-object view does not construe sentences of the form '... explains ...' as relating an explainer and an object, so he need not construe sentences of the form '... is being used to express question (provide description, describe event) ...' as relating a sentence and an object.

The present no-object view denies assumption IIA but retains analogues of I and IIB. It permits a wide range of inferences sanctioned by the question view by taking into account differences in questions expressed using different interrogative sentences and using different emphases. In return for its ontological simplification, however, there is a semantical complication. According to the question view (with the parenthesized material), 'explains' in restructured explanatory sentences is a relational predicate applicable to an explainer with respect to a question. According to the no-object view 'explains' in restructured explanatory sentences is an operator that operates on interrogative sentences to form a predicate. On the former view there is one predicate 'explains', while on the latter view there are (possibly) non-denumerably many semantically unrelated predicates containing 'explains'.[24]

[24] Davidson would take this as a sufficient reason for rejecting the view, since he holds that a language with infinitely many primitives is unlearnable.

There are consequences of this difference that might be pointed to as being more favourable to an object view. Consider the sentences

15   Plato explained why Socrates died
16   Smith explained why Jones got indigestion.

Intuitively, we might say that the states of affairs described in (15) and (16) have something in common, viz. someone's having explained something. On an object view this is reflected in the inferences which that view permits from both (15) and (16) to

17   $(\exists x)\,(\exists y)\,(x$ explained $y)$,

inferences that are precluded on no-object views. Again, from (15) and

Plato explained why he believed the theory of forms

we can infer

Plato explained (at least) two things,

an inference readily justified on the question view but not on the present no-object view. Accordingly, even though the present no-object view is able to handle inferences that stymied some of the earlier object views, there are some rather plausible inferences for which it seems to offer no justification.

In section 8 three implications of the question view for the D–N model of explanation were considered. There are analogous implications if the present no-object view is adopted. Let me consider only one. On the D–N model an explanatory sentence '*A* explains . . .' is true if *A* is a set of sentences containing laws and initial condition sentences and *A* entails a sentence describing the event referred to by an expression replacing the blank. But, letting $q\,(e, d)$ be an interrogative sentence used by a speaker to express a question about event *e* under description *d*, and *A* be a sentence or set of sentences, the following cannot be accepted:

18   '*A* explains *q* (*e*, *d*)' is true if *A* contains laws and initial
      condition sentences and *A* entails D, where *D* is a descrip-
      tive sentence used by a speaker in providing description
      d of event *e*.

Let   *e* = the event of Socrates' death
             d = ['Socrates died']$_e$
             D = the descriptive sentence 'Socrates died'
             q_1 (*e*, *d*) = the interrogative sentence (used by a speaker
                      to express a question about *e* under *d*) 'why did
                      Socrates die?'
             q_2 (*e*, *d*) = the interrogative sentence 'how did Socrates
                      die?'
             q_3 (*e*, *d*) = the interrogative sentence 'where did Socrates
                      die?'

For some explanans *A* let us assume that '*A* explains q_1 (*e*, *d*)'
is true, where *A* contains laws and initial condition sentences
and *A* entails *D*. But on the basis of earlier criteria we cannot
conclude from this that '*A* explains q_2 (*e*, *d*)' and '*A* explains
q_3 (*e*, *d*)' are true. That is, from '*A* explains why Socrates
died' we cannot infer '*A* explains how Socrates died' or '*A*
explains where Socrates died'. In short, on the present no-
object view (18) would be unacceptable as a D–N requirement
for an explanans.[25]

Comment

BY MARY HESSE

It would strange if every occurrence of 'explains' in the
language could be captured by the same formal explication,
even if this is as general as the schema subject-relation-object
proposed by Professor Achinstein. So I must confess that I

[25] I am indebted to Stephen Barker, Robert Cummins, and George
Wilson for very helpful discussions of the issues. This work was
supported by the U.S. National Science Foundation.

do not find myself in basic agreement with the presuppositions of Achinstein's paper, which, if I have understood it rightly, is just such an attempt to find a general expression for the object-relatum of the relation 'explains', where, moreover, that relation is taken to be univocal. I shall start with some brief critical remarks about the validity of such a programme; then I shall concentrate on explanation in the natural sciences, and consider how Achinstein's arguments bear on that application; and finally I shall examine his remarks about the DN model.

1. The formal analysis (or explication) of the concept of 'explanation' began with the problem of explanation in the natural sciences, and it began with elaboration of the DN model. However, there are by now at least three widely accepted general consequences of the ensuing debate which are enough to show that this model is by no means to be taken as the 'standard model of explanation'.
These are

(i) It has grave shortcomings as an explication of the structure of explanation in natural science itself.
(ii) The question of whether it usefully applies to the notion of explanation in history, psychology, and the social sciences is a matter of current controversy.
(iii) The question of whether *any* model applying to the natural sciences is adequate also for these human sciences is also highly controversial.

The relevance of these points to the question at hand is simply that it would seem to follow *a fortiori* that it is dangerous to presuppose that any single formal schema will adequately capture all the idioms of 'explains' in a natural language. Now Achinstein may well respond that his schema is only formulated for the *object* of explanation, and that he has left sufficient licence for different kinds of explanation by requiring that the question to which the explanation is an answer should be explicitly specified, and that this specification may cover many different objectives of the explanation. To this response I would make three general counter replies. First, Achinstein seems to presuppose that the relation 'explains' itself is univocal and that the form of the object is independent of the particular model adopted for that relation. Second, although the 'explainer' is said in the first sentence of the paper to be *either* a person ('Plato') *or* a theory ('the kinetic theory'), almost all the examples on which we are later asked to exercise our linguistic intuition take the explainer to be a person. It is not prima facie obvious that this restriction does not colour our view of the proper analysis of the relation and its object. 'Newton explains' and 'Newton's theory explains' are surely likely to have a different grammar. Thirdly, and most importantly, to specify the question to which a proposed explanation is intended as an answer may only be a way of restating the problem of the analysis of 'explanation', and may not in itself do anything to illuminate that problem. To take one of Achinstein's examples, 'how Kissinger shook hands with Le Duc Tho', even without the further qualifications that he considers as to time, place, manner, in whose company, etc., is capable of many different *sorts* of answer depending on context. It may be a journalist's request for the surrounding circumstances (they had signed a treaty), a historian's request for reasons (How come ?—because of Nixon's policy of detente), it may be a request for a behavioural description, for a physico-physiological description, (perhaps Kissinger had just broken both arms), for Kissinger's intentions (to impress photographers), or for Kissinger's unconscious motives. Doubtless all these could be specified by the form of question appearing as the third term in Achinstein's object-argument, but nothing has thereby been done to clarify the notion of explanation in any of these senses. Indeed, *what* is being asked for in these cases has to be picked up from the context and not from single explanatory sentences. It is considerations like this which indicate that the development of a science of linguistics adequate for the idiomatic rules of natural language will be an exceedingly complex task, perhaps more complex than has yet been understood. I do not in the least wish to deny that such a general science may be in principle possible, and that philosophical analyses like Achinstein's are contributions towards it; I would only suggest that this task should not be confused with the more manageable problem of explicating particular types of scientific explanation which are better understood and more directly formalizable.

2. I shall therefore shamelessly retreat to the original explicatum of the concept "explanation", namely the natural sciences.

Prima facie, as Achinstein himself notes, the object of explanation here is a *sentence*. To this construal of the object of explanation in general Achinstein has one reservation and one major objection.

(i) The reservation is as follows. The explication of 'explanation' he asserts, and I agree, is such that '*A* explains "...*x*..." ' entails *A* explains "...*y*...' " where *x*, *y* are thing-variables and $x=y$. That is, the object of explanation construed as a sentence should be *thing transparent*. But Achinstein goes on, since it is not the case with all relations that '*A R* "...*x*..." ' entails *A R* "...*y*...',' " where $x=y$, (for example if $R=$ 'believes'), this property of 'explains' remains unclarified in the sentence construal of the object of explanation. Now I would first remark about this that I do not think that Achinstein himself has provided any further clarification of this point in his subsequent expansion of the object of explanation into the ordered triple (event, description, question). And secondly, I am not clear what kind of clarification Achinstein is asking for here. The only sort of clarification I can think of is one which would follow an analysis of the *relation* 'explains' itself, rather than an analysis of its object. For example, if as in the DN model '*A* explains' is something like '*A* contains a general law and initial conditions and entails', the thing transparency of the object sentence is guaranteed by logical equivalence.

In parenthesis, before going on to Achinstein's major

objection to the sentence construal, I would like to inquire how micro-explanation fares in this construal. Achinstein has previously, and I think correctly, rejected any construal of the object of explanation which permits self-explanation of the form '*A* explains *A*'. He gives this as one reason for rejecting the event construal of the object on the grounds that, for example, the state of affairs of this gas having absolute temperature 300°K cannot both be explained by the state of affairs of this gas having mean molecular kinetic energy of so many joules, and be identical with this state of affairs as would usually be claimed. But even if the event construal is adopted, I do not agree that the identity of these states of affairs is fatal to it, for the states of affairs linked by 'explains' in this case are not as Achinstein portrays them, but rather

'The state of affairs that all systems of mechanical particles have mean kinetic energy of x joules, and that this gas is such a system, and that mean kinetic energy of x joules is equivalent to a temperature of 300°K' explains 'The state of affairs of this gas having temperature 300°K'.

The relata of 'explains' are both 'events', but they are not the same event, since the explainer includes but is more general than the explanandum. Therefore Achinstein does not need to resort to 'events under a description' to avoid the difficulty of self-explanation. His move is dictated only by his other objection to the event construal, namely the non-event-transparency of 'explains'.

Moreover, the sentence construal of the object *does* satisfy the requirements of micro-explanation. Achinstein introduces 'event under a description' as an ordered-pair for the object of explanation chiefly on the grounds that it overcomes the alleged difficulty of self-explanation in the event construal. For the event description 'This gas has mean kinetic energy of x' is not equivalent to the event description 'This gas has absolute temperature 300°K'. Hence the first may explain the second without self-explanation. This is true, but it applies equally to the one-term object in the sentence construal. Hence the sentence construal shares this, the major, advantage claimed for the event-under-a-description construal, although Achinstein does not mention it as grounds for the sentence construal. But it is in any case a hollow advantage, because this is not the correct formulation of the micro-explanation in a sentence construal. In any adequate explication of 'explains' in science such an explanation would surely read something like

> 'All systems of mechanical particles satisfy Newton's laws, and if they have mean particle kinetic energy x they have absolute temperature $300°K$, and this gas is such a system and has mean particle kinetic energy x' explains 'This gas has absolute temperature $300°K$'.

One does not have to be an addict of DN explanation to require that the explanans contain general laws or something equivalent to general laws, and therefore that it is not identical with the explanandum.

I conclude from this parenthesis that Achinstein has given no substantial reasons for the ordered-pair event-under-a-description construal of the object which would not equally support a sentence construal.

(ii) Achinstein's major objection to all his first three suggestions is that they do not distinguish different types of question 'why?', 'how'?, etc., asked about the object of explanation. Now clearly a general linguistics of the concept of explanation must take account of these distinctions. However, in my self-imposed restriction to scientific explanation I question whether analysis of the object is required or is even sufficient for this job. Consider a scientific analogue of Achinstein's Kissinger example.

1   Galileo's law explains why that falling body O accelerates towards the earth.

2   Galileo's law explains how that falling body O accelerates towards the earth.

(1) may mean at least two things: (*a*) 'gives a general theory to explain' (lct us call this 'theoretically explains'), or (*b*) 'gives the immediate cause of' for example, '*O* was knocked off the table'. In either sense (1) is false. If (2) means a question that is adequately answered by 'it accelerates uniformly', (2) is true, but it is easy to think of other senses in which (2) is false. In other words, there are in Achinstein's terms at least three questions that might be involved here:

1*a*   'theoretically explains why'
1*b*   'under what conditions did'
2   'according to what law did'.

However, as I rcmarked before about the Kissinger example, elaboration of these questions is merely a way of distinguishing different senses of 'explains', and does nothing to illuminate the nature of these different senses. Indeed it is somewhat less perspicuous than identifying *relational* categories of 'theoretical explanation', 'causal explanation', 'law-like explanation', 'historical explanation', etc.

3. At the end of his papcr Achinstein considers some implications of his analysis for the DN model of explanation.

(i) He claims that this modcl docs not distinguish different questions about the same object of explanation whereas his analysis explicitly builds these questions into the object of explanation.

It sccms to me that the sense in which this is true is an unfairly restricted interpretation of the DN model. Take my last example about Galileo's law. Different aspects of the DN model can certainly be used to provide explanation of the three kinds I have mentioned, depending on whether the general theory is emphasized, or the initial conditions (as Hempel suggests are emphasized in DN explanations in history), or the low-level descriptive law. Of course, it may be that the question being askcd is none of those I have mentioned, but rather 'for what purpose did *O* accelerate?', or something else again, in which case the DN model is arguably inadequate, but this is only another way of saying that if there are purposive or functional explanations in science, the DN model may not fully capture them. We did not need an analysis of the object of explanation in Achinstein's sense to tell us this.

Objection (i) is therefore not a new type of objection to the DN model.

(ii) Next, Achinstein uses his analysis to indicate why it is that some trivial 'explanans' which satisfy the DN model are not satisfactory explanans. This is because it turns out that they do not answer the questions that are likely to be asked about the object of explanation. This is true of the examples Achinstein gives, but in relation to scientific cases not very helpful, because if we characterize a question asking for an adequate scientific theory as a 'theoretically why' question, we are not thereby helped to characterize what constitutes a good answer to such a question over and above the DN model. The problem for scientific explanation remains where it was before Achinstein's analysis, namely 'what counts as a good theory?', and this problem always did suggest that the DN model is inadequate.

(iii) Thirdly, Achinstein suggests that his analysis shows that the symmetry thesis for explanation and prediction must be rejected, since to give an answer to a question that is satisfactory in some contexts does not entail that the answer is sufficient for the explanandum to have been predicted. Here the standard replies of the symmetry theorists seem to cover his examples. First, the explanation may be *incomplete* (for example we can explain why Jones got indigestion, but not why he did so on Tuesday), in which case an analogous scientific explanation of the DN kind would also be inadequate (for example it might explain why the orbits of planets are conic sections, but not why the orbit of Mars is such and such an ellipse) Or, second, the events to be explained may be governed by statistical laws, in which case the DN model and symmetry thesis have to be modified anyway to allow explanation and prediction to be made only with some probability. There are, of course, other objections to the symmetry thesis, but they seem to be ones that are concerned with the explanatory *relation* rather than with the form of its objects. In any case, Achinstein's analysis seems to leave the DN model much where it was before.

Let me conclude with a reflection arising out of the points that have been discussed, which seems to me to show that, on any satisfactory construal of the explanation relation for natural science (and there may be different relations in different contexts), the object of explanation will be a sentence.

Any request for explanation is a request to give more information about a context which will show how different parts of it are related and what further expectations we should have of it. This is why all models reject self-explanation, and the DN model in particular requires the explanans to contain general laws licensing factual and counterfactual inferences. It follows that the relation of explanation will involve some form of inference, for without this we can have no rules for forming justified expectations when the explanation is given to us. The inferences need not be deductive, they may be some form of inductive or statistical inference recognized in the literature of science. Now inferences connect sentences, and so it seems likely that the relata of the explanation relation will be sentences however that relation is construed. To show that the DN model is not the only account of such a relation, consider how an explication of 'explanation' might be constructed for a confirmation model of scientific theory, for instance: 'A theory explains data if it is highly confirmed by the data, and if it enables further highly confirmed predictions to be made from the data'. If the relation 'confirmed' is construed in some such way as 'raises the probability of or degree of belief in', the arguments of this conditional probability function are, again, sentences, in all confirmation theories I know of. Notice that I have here assumed that the notion of 'acceptable inference' is logically prior to the notion of explanation. This seems to have been almost universally assumed, and it helps to explain why the object of explanation is typically thing transparent but not event transparent. This is simply in virtue of the logic of the inferences, whether deductive or inductive or proba-bilistic, which constitute explanation.

Comment

BY R. HARRÉ

Several of the points I want to make are already implicit in Miss Hesse's paper. However, I intend to concentrate on a range of problems which she has touched upon only briefly and to turn my attention to the cases where the explainer is a person. I hope to argue, albeit rather briefly,

1. that there is no one sort of item which an object of explana-tion must be, not because there is no such object, rather because there are literally for me at the moment, a bewildering variety of such objects:

2. that amongst them are items of the state of affairs kind. Thus, I will be concerned to rebut the objections offered in Section 2 of Achinstein's paper to the doctrine that the object of explanation can be states of affairs. Much depends on the exact form which the restructuring, to use Achinstein's term, of such common idioms as '. . . explained why . . .' is thought to take. Consider the examples:

> Plato explained why Socrates died.   (1)
> Plato explained how Socrates died.   (2)

It seems to me that for the restructured form to capture all that is conveyed so economically by such idioms as (1) and (2), it must include, in some explicit form, the indication of the mode of explanation provided by the various interrogatives that appear in the idiom. Thus (1) and (2) would seem to be adequately restructured only by some such sentences as:

> Plato explained Socrates' death by giving the reason for it.
>
> (1a)
>
> Plato explained Socrates' death by telling the manner of it.
>
> (2a)

I suppose the interrogatives can convey all this in the idiom since each typically introduces questions the answers to which would usually be expected to be reasons or causes in the case of 'why', and manners or means in the case of 'how'.

It will be evident that I have already departed from Achinstein's favoured doctrine that anything which is relevant to the kind of explanation which is wanted should appear in the object of explanation. It seems to me clear that the interrogatives and their associated questions, whatever be their relations to the original sentences, qualify the *explanans* that is required, rather than serve in any sense as the objects of explanation.

There are further unresolved ambiguities in both restructured forms to which I shall pay no further attention since I suppose they can be resolved by resort to further qualifications of the mode of explanation. For example in (2a) we are still left uncertain as to whether Plato told of the style of Socrates' passing, say bravely, or the means of it, say hemlock. Achinstein draws our attention at various points in his paper to this, but since it seems to me plain that these distinctions concern as much the mode or form of the *explanans*, as possible variety in the referents of the *explanandum*, we can leave any further discussion of these matters.

I would say then, that Achinstein's argument can be drastically curtailed if we agree that the first form of restructuring that he offers:

> 'Plato explains Socrates' death' (1c)

is inadequate. And as he himself comes on to notice, it will not serve to distinguish the sense of 'why', from 'how'; that is, will not serve to distinguish examples (1) and (2). So, the major difficulty that he examines in Section 5 and which he finds to beset most objects of explanation theories, can be dealt with very shortly, since though each of 'Plato explained why Socrates died' and 'Plato explained how Socrates died', have the same object of explanation, namely Socrates' death, if they are presumed to have different truth values then so must their restructured paraphrases. But, if the restructured paraphrases of both are:

'Plato explained Socrates' death',

they must have the same truth value according to Achinstein. However, if we refuse to admit that 'Plato explained Socrates' death', is an inadequate paraphrase of the original idioms, we get:

'Plato explained Socrates' death by giving a reason for it',

and

'Plato explained Socrates' death by telling the manner of it',

and the difference in truth values is preserved through restructuring. This seems to deal with the second of the major difficulties with the object of explanation of states of affairs or event view referred to by Achinstein in Section 2 of his paper.

The restructured form which he offers for the idioms we are considering is even less satisfactory when the interrogative included in the idiom is either 'what', 'when', or 'where'. Indeed, the restructuring which I have offered for 'how' and 'why' examples, will not do for these cases either. In the case of 'where', for example;

'Plato explained where Socrates died'   (3)

the restructured form:

'Plato explained Socrates' death by telling the place of it'

(3a)

is clearly not a paraphrase of (3). So the examination of the idiom leaves us much where we were as to the standing of alleged objects of explanation, but has at least drawn our attention to the many forms that an *explanans* might take.

By collecting up Achinstein's examples and adding some of my own, in which 'explain' takes a direct object, a modest schema does emerge by which a preliminary sorting of objects of explanation can be made by reference to modes of explanation. Having set it out, I shall briefly examine the groupings so made to see if they have anything in common within themselves. The results will be disappointing.

It has lately been fashionable to decry the once popular claim that certain interesting words had a variety of meanings. But in the case of 'explain', I think one would be justified in saying that there were two main meanings of the word, and perhaps even four such meanings. In my view, Achinstein lumps together altogether too promiscuously various senses of 'explain', and I hope to justify very briefly the claim that these are indeed senses of 'explain'.

Case 1. Is it persons who explain, or sets of sentences that explain? Both kinds of statements are countenanced. Do we find in them different senses of 'explain'? If it is the case that a difference in sense is shown by a difference in grammar, we can certainly make a case for differentiating 'Newton explained Kepler's laws', from 'Newton's theory explains Kepler's laws'. The grammatical difference shows up in the qualifications which one may make adverbially to 'explain'. For example, 'Newton explained Kepler's laws slowly, carelessly, yesterday, etc.', and none of these qualifications can properly be made to the verb in the case where it is the theory which explains. Newton's theory cannot be said to explain Kepler's laws either slowly or quickly, either yesterday or today.

Case 2. 'Explaining' can be giving the reasons for, means of, causes of, something or other. It may also be the clarification of, or the explication of something or other. My justification for believing that these are different senses of 'explain', comes from the fact that in order to make the difference clear, we call upon

CE

non-overlapping sets of synonyms. It would be convenient if each sense of explanation in Case 2 had a characteristic type of object, but counter-examples are easy to find. For example, as well as explaining propositions, policies, etc., we can in the same sense explain the steam engine.

I will draw upon the distinction made in Case 2, for a principle by which objects of explanation might be sorted.

Category A. When 'to explain' means 'to give reasons for', 'causes of', and the like, the objects of explanation are events, states of affairs, and similar sorts of entities. In general, when 'to explain' has this meaning, any referents of the *explanans* will be different from the object to be explained. Examples: 'Plato explained Socrates' death', and amongst other items that he might thus have explained, are the presence of a person, the persistence of a condition, the extinction of a light, the occurrence of a storm, the appearance of a dog (both its popping out of the background, and how it looks), the loss of a battle, the absence of reinforcements, the origin of a superstition. When the object of this sort of explaining is a course of action, then explaining it may serve also to justify it, as when I ask you to explain what you *had* done. But if I were to ask you to explain what you are about to do I would want something rather different by way of an *explanans*, and this takes me to the second sense of 'to explain' under Case 2.

Category B. When 'to explain' means to elaborate on or explicate, to make clear in a context of possible uncertainty or unclarity, typically propositions, and items of the same character as propositions, are amongst objects of explanation. In this sense of 'to explain', the very same item is both object of explanation and referent of the *explanans*. Examples: 'Plato explained the Athenean Constitution, the question put by Thrasymachus, the meaning of Socrates' reply, the method of dialectic, what Socrates was to do with the cup of hemlock,' and so on. In this sense, Harvey explained the circulation of the blood to the Court, and James Watt explained the steam engine to the shareholders of the Stockton and Darlington Railway Company.

Ideally, examination of the objects of explanation categorized in the first instance by the sense of explanation of which they are the object, ought to yield criteria which would produce the same categorization independently of the senses, but while the items of Category A are at least (i) non-linguistic and (ii) 'of the world' in some way, the items of Category B seem to defy any unification. Not only are there both linguistic and non-linguistic items (a constitution, a question, the circulation of the blood) but the items are of very diverse kinds. For example, Plato explained a particular constitution, while Harvey explained the circulation of the blood in general, not the circulation of the blood of the particular specimen he was using in the demonstration. I admit myself baffled by the heterogeneity of Category B and even more so by the two categories taken together. I suppose in consequence that there is no universal characterization of 'object of explanation', if the two senses of 'explains' are run together, and that in Category B we are hard put to find any property which the objects of explanation have in common.

I will content myself with an attempt to clarify a little the standing of the items in Category A, that is presences, losses, extinctions, origins, occurrences, appearances, persistences, and so on, in the hope of defending the view that when Plato or someone else undertakes to explain any of these items, the item so explained is the object of explanation for him.

The first point to notice about items in Category A is that they are particulars, thus one might ask for the presence of water on the moon to be explained. One could not be asked in the sense of Category A to explain persistence, but rather the persistence of Kutuzov in refusing to give battle after Borodino.

But the admission of losses and absences to the list of items in Category A makes the next step more difficult. Occurrences and presences are, I suppose, events and states of affairs, but whether one ought without more ado to include losses amongst events and absences amongst states of affairs, I think could be debated. Even so, why should we refrain from following our intuitions, and accepting the items in the above list, as at least some objects of explanation?

I have already dealt with one of Achinstein's arguments, namely that the idioms containing 'how' and 'why' are not adequately rendered by the simple-minded paraphrase which he offered initially. However, he does bring a rather stronger argument to bear against the events or states of affairs categorization of alleged objects of explanation. This argument rests on the problem created by his truth-under-substitution-preserving conditions IIA and IIB, by the possibility that different descriptions of the same event may on substitution one for another change the truth value of the explanatory sentence, and rather than countenance this he prefers to abandon the idea that there is an object of explanation. But again it looks as if the variety in the truth value stems from a variety in the explanations Plato may be presumed to have given, which I shall argue, stems from a corresponding variety in the object of explanation. It seems to me quite unexceptionable to suppose that a given event may, variously qualified, be given a variety of explanations. And further, that various descriptions of that event occurring in explanatory sentences, serve at least to make clear which explanation is being offered and what exactly is being explained. Thus diversity in modes of explanation may reflect diversity in objects of explanation, but objects of explanation may be diverse under the same mode of explanation.

One of the reasons why an event is susceptible of a variety of explanations is not because this variety stems from there being various modes of explanation, but that an event such as Socrates' death is an entity and may have properties predicated of it. Thus far Goldman is certainly right. For instance, there are Socrates' death, Socrates' quick death, Socrates' death from drinking hemlock, Socrates' tragic death, and so on. One source of variety of explanation seems to match the variety of properties by which Socrates' death is qualified, provided the existence of the properties is what is taken as calling for an explanation. But each different state of affairs may be susceptible of the same mode of explanation, that is, its cause, for example, or reason, may be sought. Socrates' death is not, incidentally, adequately expressed as Socrates having the property of having died, and in this respect at least Goldman's account needs some re-stating.

Thus, when Plato explained Socrates' death and Socrates' quick death on two successive occasions is he, or is he not, explaining the same thing? Why can't I say, if he gives the same explanation in both cases—that is, the political reason for it—then of course it is true on both occasions that he explained Socrates' death, and the rapidity of it was ignored in his explanation. But if on the second occasion he remarked that hemlock was a powerful poison, he is explaining rather the quickness of Socrates' death, a wholly different object of explanation, but an in-the-world, non-linguistic entity nevertheless, and related in an obvious way to Socrates' death.

The bare fact, as we say, of Socrates' death, can be involved as referent of the *explanandum* in various explanations:

(i) because of multiplicity in the mode of explanation, that is, whether it is by the giving of a reason or the description of the means for the occurrence, and so on. This was the point made earlier as we disposed of Achinstein's difficulty about the multiplicity of explanations demanded by the 'how' and 'why' idiom.

(ii) In disposing of Achinstein's present difficulty we are forced to enlarge our conception of the object of explanation a little further because the fact that the qualification of Socrates' death in various ways leads to different explanations under the same mode, being offered by such as Plato, leads one to suppose that in giving one of these different explanations Plato was not explaining Socrates' death *simpliciter*, but, say, the quickness of it.

Achinstein has yet another argument against the states of affairs interpretation of objects of explanation, an argument by *reductio ad absurdum*. It is worked out in terms of an example from statistical mechanics. It goes like this. The molecular energy explains the temperature of gas, but the having of this energy and the being at this temperature are identical states of affairs. And so the move to the explanation of the having of a particular molecular energy by reference to the having of that particular molecular energy is possible by substitution of expressions with the same referent, leading to an absurd self-explanation. But this example can be turned. It involves a four-way equivocation between what I have called the A sense of explains and the B sense, and between someone explaining and a set of sentences explaining, etc. Not only are there these equivocations in the example but there is a further equivocation in the meaning of a temperature. Either the temperature 300°K is a thermometer reading, in which case the state of the molecular energy explains this reading in the A sense, and there is no collapse into self-explanation, since the reading and state of the gas are not identical states of affairs: or the temperature is used to ascribe a property to the gas which is explicated, that is explained in the B sense, as being the property of having just such molecular energy, in which case the referents are the same. Self-explanation is permitted and could perhaps be defended as the minimal form of explication just as '*p* entails *p*' is the minimal form of entailment. On neither alternative does the Category A object of explanation require to be rejected. Indeed, the only case that we might make out successfully for the rejection of self-explanations is in case the explainer is a person. 'Newton explains Newton' is perhaps an objectionable form of explanation, but '*p* entails *p*' is certainly an acceptable form of entailment.

But in thus seeking to rebut the argument in its own terms, I have passed over a serious problem involved in Achinstein's choice of example. The discussion in the paper in general is of the relation between a person who explains and what he explains. It is not a discussion of the relation between an *explanans* and its *explanandum*, nor the referent of its *explanandum*, if any. For the example to be wholly relevant to the main line of argument of Achinstein's paper, it ought to take the form, 'Maxwell explained the absolute temperature of gas by . . .' Put thus, it is open to Maxwell to give a Category A explanation in case he takes 300°K to be the reading of a thermometer, or a Category B explanation in case he takes it to be a property of the gas, in the nineteenth-century sense of temperature. In the former case, the object of explanation is not identical with the referent of the *explanans*. In the latter case it is. But in neither case is collapse into objectional self-explanation possible.

I am not at all clear how an account of the way a person explains something is related to or can throw light upon the way an *explanans* explains something. But it is an interesting problem and should be pursued. The supposition that either the verb 'to explain' is univocal in this dimension, or that the transition from a person explaining, to sentences explaining, is unproblematic, is certainly arguable. It is assumed not only in this discussion, but at the end of Achinstein's paper, where, in a discussion of the no-object view which was couched in its initial stages in terms of Plato-explained, the investigation of the effect of that view on D–N theories of explanation, the explainer suddenly becomes a set of sentences.

So, by careful navigation between the two main modes of explanation possible for objects of explanation in general, derivative from the distinction between 'how' and 'why' questions, and by distinguishing between explicating and giving a reason, we can avoid Achinstein's conclusion that 'rather than denying the truth of the theoretical identity, the more plausible line would be to admit that a restructured explanatory sentence does not relate an explainer and a state of affairs or event'. I conclude that we are still able to entertain the view that when Plato explains Socrates' death, Harvey the circulation of the blood, and Watt the steam engine, they are severally related to an object, namely, an event, a process, and a thing.

Finally, I would like to make a remark or two about the no-object view. The no-object view can be set out fairly simply. The logical form of the example 'Plato explained why Socrates died' is given according to that view, by conceiving of an 'explains' operation on an interrogative sentence, thus:

. . . explains (why did Socrates die?)

which yields the predicate:

explained why Socrates died,

which is then predicated of Plato. So 'explains' is an operator 'E' which operates upon an interrogative sentence q to generate the predicate Eq ( ), and the existential quantification of this propositional function yields $\exists x \, Eq \, (x)$.

But the only sense of 'explains' for which the above operation makes sense is that of Category B, i.e. to elaborate, or clarify, or explicate, and under this interpretation, 'explains "Why did S die?"' does not yield 'explained why S died'. However, were the operator to be 'answers' then 'answers "why did S die?"' would yield the predicate 'explained why S died'. But to elucidate the logical syntax of the answers operator is no help in explaining 'explains'.

However, though I do not think the no-object view explication of the logical form of Achinstein's examples is sound, I do think that there are genuine cases of explanatory sentences, as Achinstein calls them, which are properly to be construed on no-object lines, but they are not of the form of the main examples discussed in Achinstein's paper. To identify them I offer a slightly more elaborate, but still modest and tentative schema, to sum up our intuitions as they have been so far revealed.

Explanatory sentences can be of four distinct forms.

1. The grammatical subject is a person, the explanation given is Category A, that is, the *explanans* has a different referent from the *explanandum*, and the *explanans* is neither a part of the sentence nor referred to by a part of the sentence. For example, 'Plato explained why Thrasymachus was there,' that is, Plato explained the presence of Thrasymachus by giving a reason for it.

2. The grammatical subject is a person, the explanation is of Category B form, and so the *explanans* has the same referent as the *explanandum*, since it is a clarification of it, but again the *explanans* is neither a part of the sentence nor referred to by a part of the sentence. For example: 'Plato explained Meno's question', that is, clarified it.

3. The grammatical subject refers to a set of sentences, the explanation is of Category A form, but the *explanans* is referred to explicitly in the sentence, since the grammatical subject term is the name of the *explanans*. For example, 'Drude's theory explains the relation between electrical and thermal conductivity', that is includes the reason for that relation being what it is.

4. The grammatical subject-phrase is a description of some state of affairs. The explanation is of Category B form and the grammatical subject-phrase is literally the explication of the phrase which is the object-phrase. For example, 'The molecular and kinetic energy of the gas explains the absolute temperature of the gas'. In this last form we arrive at a genuine case of no-object. In the example, both the *explanans* and the *explanandum* appear in the sentence as parts of it, and since the explaining is nothing but the clarification of the *explanadum*, it—the *explanadum*—is what has been explained, not what it and the *explanans* both refer to, namely the state of the gas. Thus there is no object of explanation though something has indeed been explained. I conclude, then, that there are rather special cases where explaining has no object, but that in many cases the object of explanation is just what we take it to be.

In sum then, my differences with Achinstein derive from the tendency which I detect in his argument to lump everything which has to do with the clarification of what is to be explained into the object of explanation. Two matters have to be clarified before we can explain anything. The problem is how they are conveyed and how the means of their conveyance is related to the object of explanation. These two matters derive from the two-way multiplicity that infects explanation. There is the multiplicity of modes of explanation, and interrogatives are not wholly adequate to differentiate the modes since, for example, the answer to 'Where did Socrates die?' is not an explanation of his death. Then there is a multiplicity of the objects of explanation, reduced in the first instance, by the distinction between giving a reason for, or cause of, and giving a clarification of something, but the definitive identification of the object of explanation is created by the possibility that the adverbial qualification of verbs of action, the adjectival qualification of nouns and noun-phrases, serve to identify one from among the variety of in-the-world objects clustered round a single event, as for instance, around Socrates' death there are, in the world, the quickness of Socrates' death, the tragedy of Socrates' death, and so on. Now there can be little doubt that the form of question 'how' or 'why', enables us to decide what is the mode of explanation. But since the form of question bears directly upon the mode and only indirectly on the object of explanation, the question ought not to form part of the object of explanation. And since the details of the description enable us to pick out exactly what is the object of explanation, the description ought not to form part of that object itself, either.

Reply to Comments

BY PETER ACHINSTEIN

1. Since both of my critics claim that 'explain' has more than one meaning I shall begin with this point. Although I do make the assumption in the paper that the meaning of 'explain' does not change with a change in object or explainer (see footnote 1), such an assumption need not be made if one wants to hold an object-of-explanation view. That is, one can be committed to theses of forms A or B (section 1)—or even to stronger theses—without this assumption.

Consider the explanatory sentences

1   Plato explained Socrates' death
2   Plato explained his theory of forms,

and assume that 'explained' has a different meaning in each sentence. One can still hold a strong version of the question view, for example—one satisfying (10) of section 7—and maintain that (1) and (2) are paraphrasable, respectively, into

3   Plato explained why Socrates died
4   Plato explained what his theory of forms says,

both of which are restructured. In short, one can hold the strong view that any explanatory sentence whatever is paraphrasable into a restructured one whose object term refers to a question, without needing to assume that 'explain' has the same meaning in all explanatory sentences.

To be sure, if 'explained' in (1) and (2) has different meanings one is precluded from drawing certain inferences. For example, on a 'monistic' view, from (3) and (4) it is possible to infer

5   Plato explained at least two things,

an inference precluded on the 'pluralistic' view. But the loss of this inference does not strike at the jugular of the object-of-explanation position, so long as (5) is inferable from suitable pairs of explanatory sentences in which 'explained' has the same meaning.

I do think that the monistic view of explanation is plausible, at least as far as sentences such as (1) and (2) are concerned. However, I cannot pursue the question here (but see my *Law and Explanation*, ch. 4). I might add that I don't find one of Harré's arguments for the pluralistic view convincing. Harré claims that there is a grammatical difference between

6   Newton explained Kepler's laws

and

7   Newton's theory explains Kepler's laws

in virtue of the fact that Newton but not Newton's theory can be said to have explained Kepler's laws either slowly or quickly, either yesterday or today. He concludes that this grammatical difference entails a difference in the sense of 'explain.'[1] I don't find this argument convincing, since by parity of reasoning we could claim that there is a grammatical difference between

8   Newton fell to the ground

and

9   Newton's apple fell to the ground

in virtue of the fact that Newton but not his apple could be said to have fallen to the ground either with or without the intention of doing so, either before or after discovering gravity. Shall we conclude that 'fell to the ground' has different senses in (8) and (9)? Perhaps 'explained' does have different senses in (6) and (7), but if so Harré's argument does not show this.

Equally, I don't find Miss Hesse's reasons for pluralism convincing. She claims that a sentence such as

10   Galileo explained why that falling body accelerates towards the earth[2]

may mean at least two things:

11   Galileo gave a general theory to explain, etc.
12   Galileo gave the immediate cause of, etc.

And she concludes that these are different senses of 'explain.'

It seems to me that there is another way of describing this situation which is at least as plausible as Miss Hesse's (and I think more so). This is to say that in doing what is described in (10) Galileo might have been doing what is described in (11) or (12), i.e. he might have explained the acceleration in

---

[1] Miss Hesse at one point also suggests that grammatical differences between 'Newton explains' and 'Newton's theory explains' will make a monistic view dubious.

[2] I am changing her example slightly to better express the point I want to make.

different ways. He might have given two different explanations. On this construal we can say

13   Galileo explained why that falling body accelerates towards the earth either by giving a general theory or by giving an immediate cause

without having to treat it like

14   He fired his rifle and his secretary.

Perhaps there are two senses of 'explained' operating in (13)— as there are two senses of 'fired' in (14)—but if so Miss Hesse has not shown this. Nor is the fact that either (11) or (12) may make (10) true sufficient to show this.

    2. I turn now to Harré's very interesting proposal that as restructured paraphrases for an explanatory sentence such as (1) above we take sentences like

15   Plato explained Socrates' death by giving a reason for it
16   Plato explained Socrates' death by telling the manner of it.

On this proposal, as I understand it, we are to treat 'explained' as a non-isolatable part of longer predicates of the form

17   explained . . . by giving a reason for . . .
18   explained . . . by telling the manner of . . . ,

in which the blank is to be replaced by terms purporting to refer to the object of explanation. Accordingly, if (1) is restructured as either (15) or (16), then we can say that Socrates' death (an event) is the object of explanation in (15) and (16), and hence in (1). But we avoid the difficulties with the event view that I mention in section 2. That is, if we suppose for the sake of argument that (15) is true but (16) is false, then unlike the event view discussed in section 2, we do not have here a violation of the referential assumption IIB (section 1). It is not the case that (16) can be obtained from (15), or vice versa, by substituting co-referring explainer-terms or object-terms. Therefore assumption IIB does not require that (15) and (16) have the same truth-value.

Harré's proposal bears affinities to what I have called no-object views (except that he now calls the objects named by terms replacing the blanks in (17) and (18) objects of explanation). And it has similar semantical complications of such views. On his proposal there will be semantically unrelated predicates containing 'explains'. So, for example, on such a view we cannot infer

5   Plato explained at least two things

from (16) and

19   Plato explained Socrates' trial by giving a reason for it.

Nor can Harré reply that this inference should be blocked anyway since the meaning of 'explained' differs in (16) and (19). On his view, as I understand it, we cannot say that the meaning of 'explained' differs in (16) and (19) since 'explained' is not an isolatable predicate with an isolatable meaning in (16) and (19) but rather an unisolatable part of a longer predicate. (Indeed, even if Harré did not adopt the proposal to paraphrase a sentence like (1) into a sentence like (16) and he agreed that explained' is an isolatable predicate in (16) and (19), on his own dualistic view the meaning of 'explained' is not different in (16) and (19). Both (16) and (19) appear to fall into his category A.)

An advantage of the object-of-explanation views that I consider—an advantage not shared by Harré's proposal—is that such views do sanction inferences such as the one from (16) and (19) to (5). This is because on such views 'explained' in (16) and (19) is an isolatable predicate relating an explainer and an object. To be sure, on such views (16) and (19) would need to be restructured. For example on the question view (16) and (19) might be restructured respectively as

20   Plato explained what the manner of Socrates' death was

21 Plato explained what the reason for Socrates' trial was.

And from (20) and (21) we derive (5).[3]

There is another reservation I have with Harré's proposal. It faces a difficulty similar to that faced by the event view in cases of micro-explanation. Consider

22 Jones explained *this gas' having a temperature of 300°K* by giving a molecular description of that state of affairs (e.g., by saying that the gas has a mean molecular kinetic energy of 6.21×10^{-21} joule),

which, I shall assume, is restructured on Harré's view. (The italicized expression denotes that state of affairs which is the object of explanation.) Now if we assume that

23 the state of affairs of this gas' having a temperature of $300°K$ = the state of affairs of this gas' having a mean molecular kinetic energy of 6.21×10^{-21} joule,

then, on Harré's proposal, from (22) and (23) we should be able to infer

24 Jones explained this gas' having a mean molecular energy of 6.21×10^{-21} joule by giving a molecular kinetic description of that state of affairs.

The trouble comes when we suppose, as well we might, that (22) and (23) are true but (24) false.

3. While I am on the subject of micro-explanations, let me comment on my two critics' remarks on my discussion of this

[3] Earlier I said that even if it were true (which I don't think it is) that the meaning of 'explains' is different in (3) and (4), so that (5) can't be inferred from (3) and (4), this would not strike at the jugular of the object-of-explanation position. Now I am saying that it is a decided advantage of the question view over Harré's object view that (5) can be derived from (20) and (21). These are not inconsistent claims, since even if one wants to be a dualist and hold that the meaning of 'explains' is different in (3) and (4), it is, I think, less plausible to be a dualist about 'explains' in (20) and (21). (Harré, if I understand him, is a monist with respect to these sentences.) And if one is a monist here then the inference from (20) and (21) to (5) is quite plausible.

subject in section 2. Miss Hesse wants to say that this gas' having a mean molecular kinetic energy K does not explain its having a temperature T. Rather the explainer, even if construed as a state of affairs, is more complex, since it must include a law. I don't quite understand her formulation of this more complex state of affairs, so let me try to reformulate it.

Consider the following sentences

(*a*) Any gas whose mean molecular kinetic energy is K has a
      temperature T (law)
(*b*) This gas has a mean molecular kinetic energy K
(*c*) This gas has a temperature T.

Perhaps Miss Hesse's point is that if a state of affairs is taken to be the explainer then we should say this:

25   It is the state of affairs described by (*a*) and (*b*) (and not
     simply by (*b*)) that explains the state of affairs described
     by (*c*).

However, the problem of self-explanation for the non-linguistic view will also arise in this case. If states of affairs are objects of explanation, then since (let us suppose) the state of affairs described by (*b*) = the state of affairs described by (*c*), from (25) we may infer

26   the state of affairs described by (*a*) and (*b*) explains the
     state of affairs described by (*b*).

But (26) I would regard as self-explanation (or perhaps better, as what Hempel once called partial self-explanation), and therefore objectionable. The same problem emerges if one adopts a *sentence* view of the explainer, although I agree with Miss Hesse that it does not arise if one adopts a sentence view of the object of explanation.

Harré sees equivocation in my discussion here. A temperature of 300°K (he says) can be a thermometer reading, in which case the state of affairs of this gas' having that temperature $\neq$ the state of affairs of its having a mean molecular kinetic energy of $6 \cdot 21 \times 10^{-21}$ joule, and if so no self-explanation results.

(The explanation of the one state by appeal to the other is what Harré calls an *A*-explanation.) Or a temperature of 300°K is not a thermometer reading but a property of the gas which is *B*-explained ('explicated') by the gas' having a mean molecular kinetic energy of 6.21×10^{-21} joule, in which case the states of affairs in question are identical. But, says Harré, where *B*-explanation is concerned self-explanation is permitted and can be defended as a minimal form of explanation, just as '*p* entails *p*' is a minimal form of entailment.

Even if we were to grant Harré's distinction between two senses of explanation I would not grant his claim that the *B*-sense allows self-explanation. I don't see how it is possible to explicate, elaborate, or make clear—Harré's terms for *B*-explanation—the state of affairs of this gas' having a mean molecular kinetic energy *K* simply by appeal to its having a mean molecular kinetic energy *K*. This does not seem to me to be even a minimal case of explanation. Nor, again, granting Harré's distinction between two senses of 'explain' is it obvious to me that when we explain why this gas has a temperature *T* by appeal to its having a mean molecular kinetic energy *K* we must construe this as a *B*-explanation if temperature *T* is not construed simply as a thermometer reading. In classical thermodynamics the temperature of a gas is one of its thermodynamic properties (like pressure, volume, and entropy); it is not simply a thermometer reading. Yet when appeal is made to statistical mechanics in explaining why bodies exhibit the thermodynamic properties and relationships they do this need not be construed simply as explicating, elaborating, or clarifying the claims of classical thermodynamics.

4. Finally, I shall briefly comment on a number of unrelated claims of my critics.

(*a*) Miss Hesse writes: 'Prima facie, as Achinstein himself notes, the object of explanation here [in the natural sciences] is a *sentence*.' I'm afraid I made no such claim. I do speak of this view as the most widespread *philosophical* view, but if anything is prima facie about objects of explanation I would say that it is that they include (in addition to sentences) events, states of affairs, phenomena, facts, etc.—even in the sciences.

(*b*) One of my objections to the sentence view is that it does not account for thing-transparency. Miss Hesse complains that the question view provides no clarification of this either, and she adds that such clarification can come only by analysing (the meaning of?) 'explains' itself, not its object. My reply is that the question view provides identity criteria for questions that, together with its conception of the logical form of explanatory sentences, sanctions an inference from 'Plato explained why Socrates died' and 'Socrates = Xantippe's husband' to 'Plato explained why Xantippe's husband died.' For this purpose the question view does not require an analysis of the meaning of 'explains.'

(*c*) I claimed that one who adopts the question view will have to give up the explanation-predication symmetry claim of the D–N theorists. An explanation may explain why Jones *got indigestion* on Tuesday without explaining why he did so *on Tuesday*. And I gave an explanans for the former which is such that had one known it in advance one would not have been able to predict that Jones would get indigestion on Tuesday. Miss Hesse is unconvinced because, she says, the D–N theorist could reply either by saying that the explanation is *incomplete* or else by saying that it is not deductive but *statistical*.

Neither of these replies should satisfy the question theorist. To explain why Jones *got indigestion* on Tuesday and to explain why Jones got indigestion *on Tuesday* is to explain different things, on the question view, since the questions are different. One can answer one question completely without answering the other. The trouble with the symmetry view is that it makes every question asked one in which all or most of the terms used in making it are emphasized (why did *Jones get indigestion on Tuesday*?) No doubt such questions can be asked, but so can ones in which one word or phrase is emphasized. I have not incompletely answered 'why did Jones *get indigestion* on Tuesday?' simply because I have not answered 'why did Jones get indigestion *on Tuesday*?' Nor does the answer to the former that I considered provide a statistical explanation.

(*d*) Harré claims that the only sense of 'explains' for which the no-object view makes sense is the clarification sense, since 'explains "why did Socrates die?"' can only be understood as 'clarified the question "why did Socrates die?"' In my paper it would have been better if I had said that on the no-object view "explained" in "Plato explained why Socrates died" is to be construed as an operator on an *indirect* (and not a direct) interrogative. This, I think, would meet Harré's objection, since there is no temptation in the latter case to opt for a clarification sense.

II/Teleological Explanation

Peter Geach

The question what things are *for*—what it's all in aid of, in the current phrase—is a weighty question, not to be answered just with words about words. But I am not discussing that question or that type of question; I am not giving teleological explanations, but discussing the topic of such explanations. Now explanations are given in words, so I shall need to talk about the use of words in teleological explanations and the logical structures involved. This does not mean escape into a metalanguage to avoid talk about things; words relate to things, and only in this relation can the use of words be understood. But it will be necessary to begin by turning our attention upon certain very dangerous snares that we may fall into in our ordinary teleological discourse and our philosophy about it.

The term 'teleological explanation' would quite naturally be glossed by 'explanation in terms of ends', which provokes the question: What are ends? But here I think we should already have started on a wrong tack. By a sort of linguistic original sin, human beings engaged in abstract thought have a tendency to turn constructions of a non-nominal kind, predicative phrases for example or again sub-clauses of a sentence, into noun-phrases; and then, if we are philosophers, we may ask what objects these noun-like or name-like constructions (Kotarbiński has called them 'onomatoids') correspond to *in rebus*.

To ask the question what ends are, or to construct subject-predicate sentences of the form '—is an end for so-and-so', is in my view to risk seriously muddling ourselves; to see that such questions and answers are inappropriate, it may perhaps help us to remember Quine's example: though we use, and use in talking teleologically for that matter, the construction 'for the sake of', none of us would ask what a sake is. Such sentences as 'Money is the miser's end' or 'God is man's last end' are logically misleading.

Ends at least include the things we want. Now the verb 'want' is logically an auxiliary verb like 'ought' or 'can', and a noun-object following it should be construed either as a nominalization of an infinitive or as the object of a suppressed infinitive. 'He wants power' means 'He wants to dominate over others'; 'The miser wants money' means 'The miser wants to hold and possess money'. The oddity of the philosopher's example 'She wants a saucer of mud' arises just because it is so unclear what she wants *to do with* a saucer of mud.

A similar point again arises about enjoyment, which Aquinas rightly discusses in connection with ends: a noun-object after the verb 'to enjoy' either replaces the '-ing' form of some verb (to enjoy sex is to enjoy copulating) or is the object of a suppressed verb—here we may remember Lewis Carroll's Bruno wondering what 'to enjoy plum cake' was: his sister glossed the verb 'to enjoy' in this context by 'to munch and to like to munch', so that Bruno was understandably distressed by the question, a few moments later, whether he was enjoying *himself*. All the same, 'to enjoy plum cake' does roughly mean 'to enjoy munching plum cake' (in other cases one would naturally insert another verb); and 'to enjoy munching' does mean 'to munch and to like to munch', or at least that will do for the moment as a rough account. There are complications about the notion of enjoyment that I cannot here do more than hint at. Professor Anscombe has pointed out that 'I enjoyed riding with the King' has an ambiguity that can be brought out as follows: 'I enjoyed the fact that I was riding with the King, but since I

am a very bad rider I scarcely enjoyed the experience of riding
with the King'. To understand this distinction better, we should
not ask ourselves how a fact differs from an experience; the
concepts *fact* and *experience* are what Wittgenstein would have
called formal concepts—what we clumsily try to *say* by using
these words is something that would be *shown* by an analysis
that brought out the difference of logical structure. But however this analysis may turn out, the point I have already made
will stand: 'enjoy' is logically an auxiliary verb, to enjoy is
always to enjoy F-ing.

Simple as these theses are about wanting and enjoying, they
are already controversial. What is wanted and enjoyed, someone may protest, is a thing, not something to do with a thing:
a reality, not some abstraction. In particular, it is God himself
whom the godly desire and hope to enjoy for ever, not something
to do with God. The protest is misjudged. No doubt the godly
man desires God as the hart desires the springs of water;
but after all what the hart wants is *to drink of* those springs;
and the Psalmist's wish too is expressed not just by uttering
the name of God, but in propositional style: 'When shall I
come to appear before the presence of God?'

The object of wanting and enjoying must in fact be expressed
in propositional style: we require not only a verb other than
'want' or 'enjoy', but a subject for that verb. This easily
escapes our notice for 'enjoy', since here the subject to be
supplied harks back to the subject of the main verb: it seems
superfluous to say 'I enjoy *my* munching plum cake' or 'I
enjoy *my* riding with the King'. But what is enjoyed logically
can be first desired; and that e.g. 'I want to visit Warsaw' is
short for 'I want myself to visit Warsaw' is made clear by the
contrast with 'I want Jenny to visit Warsaw': 'Jenny to visit
Warsaw' is obviously derived from the proposition 'Jenny
visits Warsaw' or 'Jenny will visit Warsaw'.

I could develop a similar point about intending. 'Intend',
or 'mean' used in that sense, is more visibly auxiliary than 'want'
and 'enjoy'—more commonly followed by an infinitive. And

again it is manifest that the underlying construction is one in which the infinitive has a subject: 'I mean to go to Oxford' is short for 'I mean *myself* to go to Oxford', in contrast with 'I mean *Jenny* to go to Oxford'. What is intended, the tradition tells us, is the end: I agree, but what the end is has to be spelled out in a propositional structure.

I have deliberately used the expression 'propositional style' and 'propositional structure' rather than the plain 'proposition', because the clause giving the content of want or intention will not always be a proposition with a definite truth-value. If Smith and Jones both want to marry Miss Brown, what each of them desires can be expressed in the clause 'I shall marry Miss Brown'. But in a monogamous society if this comes true for Smith it does not come true for Jones. All the same, it is not a mistake to say that Smith and Brown desire the same thing; the class of Miss Brown's suitors, the class of people determined by the predicable

—desires that *he himself* shall marry Miss Brown

is a well-defined class, because the predicable says the same thing about everyone whose name is inserted in subject position. It is not like the 'predicable' which we may extract from an example of Quine's:

—was so called because of his size:

it would be a bad joke to speak of a class of men including Giorgione each of whom was so called (was called *what*?!) because of his size. Here, we don't know what is predicated till we have filled in the subject of predication; but in the other case we do—clearly we know what it is to be a suitor for Miss Brown's hand, to desire *oneself* to be Miss Brown's husband, without needing to know who the suitor is, let alone what his name is.

The point I am making is of some importance in regard to the concepts: want, desire, intention. The measure and sense in which psychological egoism is true will depend on the solution to this problem. (To enjoy, I have said, is to enjoy F-ing done by *oneself*: can one desire as an end what one could not enjoy ?) It may be further brought out thus: The pronoun 'I' changes its reference with change of speaker, but in a way does not change its sense; as Frege said in *Der Gedanke*, when each of us uses the first-person pronoun there is involved a quite peculiar mode of presentation (*Art des Gegebenseins*) in which each is presented to himself alone. (This phrase is, of course, the phrase Frege used for sense, in contradistinction to reference.) The logic of 'I', or correspondingly of what grammarians call the indirect reflexive pronoun, which in indirect speech replaces the 'I' of direct speech, is a troublesome business; fortunately I shall not need to go further into it, but it is necessary to allude to it when trying to show in what sense the expression giving the content of want, intention, or enjoyment is propositional.

Even as regards wants, intentions, and desires the mode of presentation is less important than it is for thought or judgement. For, as Aquinas put it, knowledge is a matter of the way things are in him who knows, will is related to things as they are in themselves. So the mode of presentation is the heart of the matter when we are concerned with knowledge, or indeed with opinion or mere supposition; because different modes of presentation are involved, knowing that General de Gaulle was in the United States would be different from knowing that the Frenchman with the largest nose was in the United States, even if General de Gaulle was at that time the Frenchman with the largest nose. But the achievement of a state of affairs is independent of the mode of presentation: bringing it about that General de Gaulle is dead is the same thing as bringing it about that the Frenchman with the largest nose is dead if General de Gaulle is the Frenchman with the largest nose.

We are now at the place where we can consider purely teleological propositions—propositions affirming that something happened to a certain end with no reference to some desire or intention, nor yet to any subject who desires or intends.

Since there is a relative indifference to the mode of presentation even when the end figures as the object of desire or intention, we may expect this indifference to be complete in pure teleological statement; and the logical complications of attitude towards oneself anyhow vanish where nobody *whose* end the end is is even mentioned. So I maintain that the logical form of pure teleological statements is '*p* in order that it should come to pass that *q*', or for short '*p* in order that *q*'; and here we may simply take '*p*' and '*q*' to represent propositions. The complications I have just been discussing prevent us from straightforward use of the schemata 'A desires that *q*' or 'A intends that *q*' with '*q*' read as a proposition; but here there is nothing to forbid such a reading of '*q*'.

It would be a gross misunderstanding to take me to be saying that the end alleged in purely teleological propositions, or the object mentioned when we ascribe desires or intentions, is itself a Proposition. I here write the word with a capital, because what is in question is not what I call a proposition— a recognizable bit of language used in a certain way—but rather a kind of abstract entity that is supposed to be the 'meaning expressed by' propositions in my sense. The mere idea of a Proposition is to my mind a muddle; the error has old deep roots. In the *Theaetetus* Plato half grasped the idea that what is known (we may add: what is believed or even thought) cannot be given in a name but only in complex discourse, *logos*: *logos*, he was to add in the *Sophist*, has in the simplest case a subject-predicate structure. But the lesson has proved very hard to learn. If Jones knows that the Earth is round, and Smith believes that the Earth is flat, then there is a constant tendency to treat these *that* clauses as *definite descriptions* of what Jones knows and what Smith believes; we could in principle use proper names instead—Jones knows Piff and Smith believes Paff, where Piff just is the Proposition that the Earth is round and Paff just is the Proposition that the Earth is flat. To my mind this is mere nonsense: it is one more manifestation of that nominalising Original Sin that I have already mentioned. And as regards the objects of desire and will we pass from latent to patent nonsense: if Robinson wishes Julius Caesar had never been murdered, this certainly does not mean that Robinson wishes the Proposition that Julius Caesar was never murdered, or wishes Puff, if 'Puff' is the proper name of that Proposition.

I turn from this to another misunderstanding about ends, of some importance in the history of Aristotelian and Scholastic thought. Aristotle several times refers to a certain ambiguity of the Greek preposition '*heneka*', an ambiguity shared by the conventional English rendering 'for the sake of': the noun-phrase following it may relate to the benefit—'for the sake of freedom'—or to the beneficiary—'for the sake of his country'. (The word 'cause' has a corresponding ambiguity, in the quite familiar use in which it does stand for an Aristotelian final cause: 'the cause of freedom', 'the cause of his country'.) In his rather crabbed Greek, Aristotle used the genitive of the relative pronoun ('*hou*') to mark the 'benefit' sense, and the dative ('*hōi*') for the 'beneficiary' sense. All the modern translations I have seen agree that this was Aristotle's meaning, the dative being a 'dative of advantage'. One plan where he uses the distinction is in explaining the senses in which cosmic processes are for the sake of God: God, being unchangeable, cannot be a beneficiary of the cosmic processes, but he can be the object of desire. By what I have already said, the positive side of this raises difficulties: I shall return to these.

The medieval translators of Aristotle turned crabbed Greek into barbaric and unintelligible Latin: '*finis cuius*' and '*finis quo*' were made to correspond to Aristotle's '*hou*' and '*hōi*', the dative of advantage being misread as an instrumental ablative. Striving to understand this, Aquinas gives us the doctrine that the *finis cuius* is the thing enjoyed, the *finis quo* the enjoyment *by* or *in which* (hence, presumably, the ablative!) the thing is enjoyed: God is the saint's *finis cuius*, the Beatific Vision the *finis quo*; money is the miser's *finis cuius*, the acquiring and possession of money his *finis quo*. Apart from the misreading of Aristotle, this doctrine is open to exception regardless of provenance. The expression in which the *finis quo* is given is at least near to propositional style, being a nominalized form: e.g. 'the acquiring and possession of money' readily transforms into '—acquires and possesses money'. But if we try to think of this *finis cuius* as being primarily the end, that is just the way to run into the confusion I have thus far been trying to escape from. A mere term, like 'God' or 'money', does not serve to state an end: only a propositional *logos* can do that, and the *logos* into which we are meant to expand a term may be none too clear. Since the *finis cuius* is meant to be expressed in a term that is *not* got by nominalizing a proposition, I conclude that for our purpose the notion of the *finis cuius* is useless and misleading.

If this gives offence to pious ears, I can only say that to take offence is to mistake the level of discourse we are engaged in. I am not saying that it is absurd to thirst for the living God, or ridiculous even to talk that way; I am only opposing a too naïve way of parsing this sort of expression in logical grammar. Of course much of our time we do not advert to the logical grammar of what we are saying, and in principle we could not do so all the time; but we need to notice it sometimes, particularly when engaged in theology; and of course Aquinas was aware of this need—for example his articles on the Trinity in the *Summa Theologica* refer over and over again to the logical grammar of the language being used, to the logical force of genitive and ablative and prepositions.

I return then to my paradigm for purely teleological statements: '*p* in order that *q*'. Obviously this is not a truth-functional compound of '*p*' and '*q*'; all the same, we have here, I should argue, a kind of limited extensionality. If '*p*' is short for a proposition mentioning some individual, then the truth of '*p* in order that *q*' does not depend on the mode of presentation of that individual: and similarly for '*q*'. The matter is more obvious for the '*p*' end than for the '*q*' end, but the same holds for both ends. The logical role of '*q*' is strongly analogous to (indeed, connected with) the role of '*q*' in '*a* brings it about that q'; and there the limited extensionality I am asserting is easily demonstrated. Many ordinary transitive verbs, with no shadow of intentionality about them, are causative in sense: e.g. 'a kills so-and-so' means roughly 'a brings it about that so-and-so is dead' (some restriction may be needed on the manner of bringing about, but that is irrelevant for our purposes). Now if the truth-value of 'a brings it about that so-and-so is dead' were affected by the mode of presentation employed in the expression abbreviated as 'so-and-so', then so would be the truth-value of 'a kills so-and-so'; and similarly for all other transitive verbs with a causative force. This result is quite absurd.

I am of course aware that arguments have been put up to show that this notion of half-way extensionality is inconsistent: that if 'a brings it about that $F(\ldots)$' is extensional for replacements of singular designations, then 'a brings it about that p' must be extensional for replacements of 'p', i.e. only the truth-value of 'p' can matter. To deal with these arguments would be technical and tedious; I think I can detect a dubious assumption or a fallacious step in every argument of this kind that I have seen. But even if I were unable to detect a flaw, I should be confident that this was only my own lack of skill; for 'a brings it about that p' is certainly true for some true interpretations of 'p' and false for others, and on the other hand 'a boils so-and-so' or 'a brings it about that so-and-so boils' clearly is an extensional context for 'so-and-so'; so half-way extensionality just has to be recognized, and a logic that makes it impossible is simply wrong.

The point may seem narrowly logical, but I think it has deep significance. For it is in this feature of purely teleological propositions—their transparency as regards individual references—that we see, or perhaps rather as Wittgenstein might put it, that there *shows*, the direct relation of these propositions to the way things are and the way things act, rather than to aspects under which our minds consider things. It is more true of purely teleological propositions that they bear the direct relation than it is true, as Aquinas said, of our discourse regarding the will; for the teleology of will takes a certain tinge of intentionality from the understanding of alternatives that is the source of will. I think it is this direct bearing upon concrete things and processes that explains—of course it does not justify—the tendency to specify the end with a simple noun like 'God' or 'money' which is not a nominalized proposition.

If teleology, final causation, is to be taken seriously at all, then '*p* in order that *q*' must not be reducible to the form '*p* because *a* desired, intended, willed that *q*'; in that case there would not be a special sort of causation, but only a special sort of efficient cause—and this supposed kind of efficient cause would yield an explanation of events competitive on the face of it with other efficient-cause explanations. But it is possible to understand teleological propositions without this reduction. It seems quite clear that Aristotle did so understand them. For Aquinas the matter is not so obvious: but I think a benevolent reader will take his language about the *appetitus* or *inclinatio naturalis*, not as a postulation of some unobserved quasi-conative events in unconscious agents, but simply as a nominalizing mode of speech logically derivative from the '*p* in order that *q*' paradigm. If we reject the reduction just mentioned, there is no competition between affirmations of final and of efficient causality. As the Scholastic maxim says, *causae ad invicem sunt causae in diverso genere*: 'It came to pass that *p* in order that it should come to pass that *q*' and 'It came to pass that *q* because it had come to pass that *p*' nowise conflict—there is causal explanation in both directions, one way by efficient causality, the other way by final causality. The works and operations of an old-fashioned clock are a paradigm both of teleological explicability and of transparent mechanical explicability. (I am deliberately flouting the argument to be found in Mill, and copied by many from him, that we read teleology into the working of the clock only because we have abundant inductive evidence that whenever such an apparatus exists such-and-such human desires preceded its existence: for of course nobody has ever tried to collect such evidence— we just feel sure we could collect it if we wished. No doubt we could, but our insight into the teleology of the clock does not wait upon this, and we might have the same insight if unexpectedly we picked up such an apparatus of clearly non-human origin on the sands of Mars.)

My citation of a Scholastic maxim was not an appeal to authority; and I am far from thinking that Scholastic ideas about teleology were entirely sound, even in principle. I have already criticized the distinction of *finis cuius* and *finis quo*; the very word '*finis*' is a source of even more serious confusion, because of an ambiguity it shares with Greek '*telos*' and English '*end*'. It seems ridiculous to to spell this out: 'the end of a process' may mean 'the final stage of a process' or 'that for the sake of which a process took place'. But in spite of old joking sophisms about death as the end or completion (*perfectio*) of life, it is clear that the trap was not always avoided; Aristotle sometimes fell into this fallacy, and the Scholastics did so even more often. There was the less excuse for Aristotle's falling into it because he believed in the perpetuity of the astronomical and biological world-order; he certainly held that there was an answer, in the order of final causality, to the question 'Why do the heavenly bodies rotate?' or 'Why do living things reproduce their kind?'; but the 'In order that . . .' answer would certainly not be a mention of some state of affairs supposed to follow all the revolutions of the stars (say, a final perfect arrangement of them) or all the generations of living things; rather, the answer would be something like 'In order that their changing perpetuity may in its fashion participate in the unchanging living energy of God'. But for the Scholastics the movements of the stars and the generation of animals had begun in time and were to cease in time; this made the confusion about 'end' more natural, but not more excusable. I think I could produce documentary evidence that in current controversy about natural biological teleologies and the natural law this old confusion is still to be found—not just on one side; we got it too in Tennyson when he speaks of 'one far off Divine event/To which the whole creation moves', and more recently in Teilhard de Chardin's idea of an Omega-point.

One reason for labouring this point is that thereby we see how to refute a bad argument against teleological explanations in biology. If mayfly grubs hatch in order to develop into mayflies, and mayflies—which have no means of eating, only enough stored energy to reproduce—live in order to produce eggs that are to hatch into mayfly grubs—what then, it may be asked, is the whole process in aid of? Isn't it futile, like the labours of Sisyphos? The difficulty arises only because we tacitly suppose we have to look for an end *after* all the generations of mayflies; and we have no right to assume this, nor need we give an Aristotelian answer, in terms of their imitating God's eternal life, in order to see the error that in such matters Aristotle avoided.

Particular teleological explanations of processes in living things are so natural that hardly anyone avoids giving them: I have often been amused by the way that an author begins by explaining how Darwin's idea of evolution by natural selection makes teleological talk superfluous and misleading, and then after more or fewer pages steps into language ascribing purposes to 'Nature' or 'evolution'. But can our teleological thinking be systematized? I shall not pronounce on this, still less make any attempt at the task: I think I can say something positive about the way such an attempt should be carried on. I agree with Aristotle that the pattern of systematic teleological explanation will conform to the pattern of human practical reasoning, in its logical skeleton that is; so I shall now say something about the logic of practical reasoning.

On this matter there has been much discussion; the view I still hold is one that I owe in its essentials to Dr. Anthony Kenny. Part of the confusion has arisen from the bogus equation: practical reasoning = imperative reasoning = the logic of commands. Each step of this equation contaminates the purity of logical art by introducing extraneous elements: 'imperative' at least suggests A's addressing an imperative to B, and 'command' suggests even more alien ideas—authority, enforcement, and the like. We must retrace these false steps if we want our theory of practical reasoning to be relevant to teleological discourse.

The essential difference between theoretical and practical discourse is well brought out by Professor Anscombe's example of a shopping list: it is a difference in direction of fit. The role of a list that a detective draws up to record what a man purchases is different from the role of a list that the man himself uses when making purchases. If the detective finds he has a purchase on his list that the man did not make, he erases the item from the list; if the man finds he has a purchase on the list that he has not made, he makes a further purchase. Of course the item may look as little propositional as '1 doz. eggs' or '1 lb figs', but this applies indifferently to the purchaser's list and the detective's; in either case we may imagine the item expanded to say explicitly who makes the purchase in what shop. With this expansion, the items in the detective's list will be propositions; the item in the purchaser's list I propose to call *directives*. I use this word in preference to 'imperative' and 'command' for the same sort of motive as makes me use, and recommend the use of, the term 'proposition' in its medieval sense rather than 'statement': namely, I wish to avoid logically irrelevant and misleading suggestions. As Anscombe has pointed out, the relation of the directives in the shopping list to the man's actions is just the same whether the list expresses his own intentions or his wife's behests.

Aristotle holds that premises in practical reasoning state what a man's ends and wishes are; these are his starting-points, *archai*—'principles' in moral philosophy comes from the Latin translation of this, though the idea has been rather transformed over the centuries. Now let us suppose that a man accepts a set of directives A, B, C, . . . formulating his ends, and would infer from these what he is to do. (Let us also suppose—to make things easier—that his ends are all consistently realizable in this hard world!) Then the procedure of practical reasoning that suits this case may be described as follows.

Suppose that D is a feasible directive whose fulfilment will secure the fulfilment of at least one of the directives the count started with, $A, B, C, \ldots$ and will not be inconsistent with any of them. Then it is reasonable to add D to the directives the agent is to accept. The procedure is repeated with the derivation of a further directive E from the augmented set $D, A, B, C, \ldots$. And so on until a directive is derived upon which the agent is in a position to act.

Thus far I have merely been making explicit the Aristotelian account of practical reasoning. In some ways it is parallel to theoretical reasoning: one conclusion at each step is derived from a set of premises, not necessarily from just one premise; and again at each stage the set of available premises is augmented by the conclusion just drawn—this corresponds to the famous Peripatetic 'synthetic theorem' in theoretical reasoning. But there are essential differences between theoretical and practical reasoning: so far as I know Dr. Kenny was the first who drew attention to these.

(i) In theoretical reasoning it cannot be equally justifiable to pass from premises $A, B, C, \ldots$ to a conclusion D and to an incompatible conclusion D'—unless indeed the premises form an inconsistent set, and then we ought not to have accepted them all anyhow. But in practical reasoning D may be a directive expressing one way of getting our ends, and D' may be another directive expressing an incompatible way; in that case, if a man accepts $A, B, C, \ldots$ as his practical *archai*, it may be simply *up to him* whether he accepts D or rather D' as a guide to his actions. (This point is tremendously important: it is here—as I think Aquinas saw—that we find the root of free choice in God and man.)

(ii) (This is another way of bringing out the same difference.) Imagine a conspirator who has to secure that Jones shall not attend a certain meeting, and then imagine a detective who after

DE

the event has to trace Jones's actual movements. The conspirator reasons as follows: 'Jones mustn't go to the meeting. If he misses the 5.30 he cannot go to the meeting: making him miss the 5.30 wouldn't clash with any of our other plans; so Jones must be stopped from catching the 5.30.' The detective reasons as follows: 'Jones didn't go to the meeting. If he missed the 5.30 he could not attend the meeting; assuming that he missed the 5.30 wouldn't conflict with any of the other information I have; *ergo*, Jones must have been prevented from catching the 5.30.' If we recall that the role of premise will here be played by the conspirators' other plans on the one side and by the detective's other information on the other side, we see that the two cases would be quite parallel, if only the logic of directives were isomorphic with the logic of propositions. But in fact the conspirator's reasoning is sensible and calculated to win his chief's approval; whereas the detective's reasoning is fatuous, and if he reported that Jones was stopped from catching the 5.30 on the strength of the reasoning, his chief might well send him back to traffic control.

(iii) An added premise can never invalidate a piece of theoretical reasoning: what follows from a set of premises still follows if the premises are added to. But practical reasoning can become invalid from an addition to the stock of premises; for the added premise will express a new end to be achieved, and a policy reasonably inferable from the smaller set of premises—in that it secures fulfilment for some of the ends then expressed and is not incompatible with any—may be incompatible with the end expressed in the new premise. In this way practical reasoning, unlike theoretical reasoning, is as lawyers say *defeasible*. This defeasibility might seem to threaten the application of the 'synthetic theorem' in practical reasoning—threaten our right to construct a larger and larger net of reasoning by always adding the conclusion last reached to the stock of available premises. I have, however, formally shown elsewhere that no such threat exists.

A curious view has been stated that these differences between theoretical and practical reasoning arise not from a difference in logic but from an attempt to draw the wrong parallelism: practical reasoning, it has been argued, is like a theoretical process of thought in which we seek premises to yield given conclusions! Historically this came from an unhappy use of the metaphor 'mirror-image' in Kenny's classical paper to describe the relation of the two logics; as I pointed out long ago, this was a mistake. Some people may still not have taken my point, so I shall state it once more.

In reasoning we pass, not (in general) from just one premise to a conclusion, but from a set of premises; and a conclusion may follow from the set when it does not follow from any smaller sub-set or any one member. (Some logicians, like Kneale and Carnap, have felt discomfort at this asymmetrical character of certain rules of logic, and have tried to replace rules for deriving one conclusion from many premises by rules for developing one *set* of propositions from another. This may be a usable technical device, but it does not clarify the nature of reasoning; on the contrary, the desire to remove this asymmetry between premises and conclusion seems to me that desire of symmetry and system rather than truth which Moore damned in *Principia Ethica*.) If practical reasoning did indeed correspond to the transition from a conclusion to premises that would yield it, then there would be rules in practical reasoning whereby one *premise* yielded a plural set of *conclusions* but did not yield any one member of that set! Of course nothing in Kenny's or my own account of practical reasoning leads to any result like this; and so this explanation of why theoretical and practical reasoning are logically different may be dismissed.

The defeasibility of practical reasoning by added premises of course greatly adds to the difficulty of making practical reasonings into a comprehensive system; and this difficulty would hold also for teleological explanations linked together in Aristotle's way, with hypothesized ends of nature as the first premises. In particular small regions, it may be admitted, teleological ways of thinking have heuristic value. Harvey, who thought teleologically, understood the circulation of the blood far better than Descartes, who methodologically excluded final causes from natural science. Very recently Professor J. Z. Young discovered the way the pineal gland works by proceeding on the assumption that evolution would not have let the organ survive if it had no work to do. (It is ironical to compare this idea of evolution with the frequently repeated Darwinian argument that since many organs have no visible work to do we may presume they do no work and may dismiss the teleological pattern of explanation.) But for a comprehensive theory do we not need the hypothetico-deductive method, which proceeds by the well-known rule of propositional inference, without any nasty business of defeasibility by added premises?

I have to say briefly that I think philosophers' ideas of how far natural science is deductive are largely mistaken. Newton's Principia has a theory form comparable to Euclid's *Elements*; but if one takes the pains to go into detail it becomes apparent that this is only a literary form, not a real resemblance of structure. And no text-book of physics today remotely resembles Euclid, in form or content. The more fundamental the physics, the less the resemblance: nobody is going to elaborate a deductive system that may be overthrown by a result from Oak Ridge tomorrow. Woodger indeed did write a treatise on biology in the style of the other *Principia*, Russell and Whitehead's; I have never read this, but I know some distinguished logicians think well of it. Do biologists take it seriously, though?

What scientists or mathematicians do in constructing theories is not necessarily what they ought to do; as Frege said, bad ways of reasoning cannot be accredited simply because they have the cachet of fashion. I am not appealing to the practices of scientists as normative: I am answering those who suppose the hypothetico-deductive method to be accredited by the practice of scientists. Natural science in fact is not organized in Euclid-like deductive systems.

There is a reason of principle why this is so: defeasibility is a feature of reasonings that relate to *efficient* causality. In geometrical reasoning there can be no defeasible conclusions: if from a figure's having characteristics P, Q, R, it follows that the figure has characteristic S, then no added characteristics of the figure can defeat this conclusion. But if from the action of causes X, Y, Z, there would follow an effect E, the following may be nullified if the further causal factor W is present. This solves the error of those thinkers who like Spinoza and McTaggart have sought to assimilate the following of effects upon causes to entailment. Prevention and interference are notions extraordinarily neglected in accounts of efficient causality. If we take them seriously we may be preserved from the Rationalistic error just mentioned, and also from the Humean error of thinking that causality is a matter of invariable succession. Because of interference and prevention, true causal laws do not state what *de facto* always happens, but only what happens *if* nothing interferes—and that is quite a different matter. And here, as Mill recognized, we have to bring in the notion of tendency. What laws tell us is what tendencies will exist when certain agencies operate; what actually happens is the resultant—in a way that may in some cases be exactly specifiable, say by some law of composition—of the operating tendencies.

But now we see that at the root of things teleology creeps in: efficient causality itself cannot be adequately described without the idea of tendency, which is a teleological notion. When Aquinas said that every agent acts for an end, he was not rashly generalizing from the teleologies he thought he could see in natural unconscious agents; when he gave the reason that if agents didn't act for an end they would not do this rather than that, the point he was making may be put into the linguistic idiom of our time somewhat like this: There is no coherent way of describing an agent as having characteristic behaviour unless we describe it as acting in order that so-and-so may happen. Science is concerned, as Aristotle would say, with what *bouletai einai*, rather than just with what happens—

one damned thing after another. And to get what actually happens out of the conflicting tendencies of agents, we need what Austin Farrer by a happy metaphor called casuistry: the same sort of skill as in practical matters resolves conflicts of ends. We cannot avoid teleological thinking: we had better be conscious that we are engaged in it, or we shall mismanage it.

I end by trying to remove two prejudices. One is that teleological thinking is to be avoided because it uses nasty non-extensional logic. Well, we cannot avoid that either; but anyhow the objection is misconceived. The half-way extensionality of teleological propositions comes about from the connection between teleological reasoning and practical reasoning, reasoning with directives as premises and conclusion; and the logic of directives is not intentional, even though there have been attempts to prove the contrary. Secondly, people may suspect me of trying to smuggle back the will and purposes of God into scientific explanation. That was not my aim; the will of God is a most difficult topic, and the admission of a teleological scheme of explanation does not lessen the difficulties. If God is the changeless God of classical theology—and I should argue that any other conception of God is a non-starter—then God can have no end in this world, nothing to gain or lose by what happens in this world. If God is almighty, he can get anything he wants by the thoughts of his mind, so there cannot be ascribed to him any adaptation of means to ends. The teleologies of living things are wonderful, and often terrible too: but however wonderful the teleology (say) of an organism's reproduction, it cannot have been necessary for God to secure the end that way, when he can make stones into children of Abraham just by saying 'So be it!'. As Aquinas put it: God *non propter hoc vult hoc*, God is not caused to will the means by willing the end. No doubt, if there is a God, God *vult hoc esse propter hoc*—that is, will a universe such that teleological propositions '*p* in order that *q*' describe its inherent order. But God's will has no cause or reason: particularly not the idea of a best possible world, for that is an incoherent idea like the idea of a biggest natural number. Explanations come to an end somewhere and to mention God's will is to leave off explaining. This may be the *right* place to stop: we may remember the story of the Tsar who was not content to learn that a man was *always* posted to guard a lawn, but found an explanation in the long-past caprice of a Tsaritsa who wanted a snowdrop not to be trampled. Not for Tsars and Tsaritsas alone, as Divine Right theorists fancied, but certainly for them among other men, it is written: 'I said, ye are gods': man's will also is ultimate in explanation— such an *archē*, Aristotle said, is man.

Comment: Geach on Teleological Explanation

BY PETER WINCH

Professor Geach's main interest is in what he calls 'purely teleological propositions', which he thinks of as 'affirming that something happened to a certain end with no reference to some desire or intention and none to any subject who desires or intends'. A considerable part of his paper is, nevertheless, devoted to discussing issues which arise concerning the ends and actions of agents. Some of these issues play an important part in his exposition of pure teleology if only negatively, in order to point a contrast with what I shall call 'agent teleology'. But Geach also raises some interesting questions about agent teleology as such which do not enter centrally into his main argument. These questions are important in themselves and I shall follow Geach in devoting some space to discussing them before I turn to some aspects of his treatment of pure teleology.

My first question concerns Geach's contention that the 'object' of desiring or intending, though it must be expressed 'in a propositional structure' or 'in propositional style', will not always be expressible in 'a proposition with a definite truth value'.

He is thinking of those numerous cases in which a man wants or intends something *for himself*, that he himself shall do or enjoy something. (I will return shortly to the important question whether these are *merely* numerous cases, or whether *all* cases of wanting or intending come down to something like this.)

Geach's argument goes like this: Suppose that Smith and Jones each wants to marry Miss Brown. Then each can express what he wants in the words, 'I want (myself) to marry Miss Brown'. But if Smith gets what he wants, Jones does not get what he wants, and *vice versa*. So we cannot answer the question whether, say, Smith gets what he wants by asking whether the proposition 'I marry Miss Brown' is true or false since, if these words are taken as expressing indifferently the objects of Smith's and Jones' wants, they cannot be taken to 'have a definite truth value'. At the same time Geach wants to insist that we must be able to speak of Smith and Jones as each wanting the same thing, that the expression 'desires that he himself should marry Miss Brown' determines a well-defined class and that 'we know what it is to be a suitor for Miss Brown's hand, to desire *oneself* to be Miss Brown's husband, without needing to know who the suitor is, let alone what his name is'.

Now Geach does not wish to rest his case simply on the linguistic fact that we may sometimes speak in such circumstances of two people as 'wanting the same thing'. If that were all, it might be sufficient to point out that we can *also*, by way of contrasting the situation in question with one in which both Smith and Jones want some third party, Green, to marry Miss Brown, speak of them here as wanting different things. He claims that the difficulty is a more fundamental, logical, one, deriving from peculiarities in the use of the first-personal pronoun. I do not have anything to contribute to the solution of this difficulty, but I should like to point out that Geach's own move—that of denying that the objects of Smith's and Jones's wants can be expressed in a proposition with a definite truth value—creates severe logical difficulties of its own. For it seems to make it impossible to specify what state of affairs would count as the satisfaction or fulfilment of the desires in question. If Smith and Jones each wants to marry Miss Brown, then there is a possible fact the obtaining of which constitutes the fulfilment of Smith's desire and another possible fact (incompatible with the first one) the obtaining of which constitutes the fulfilment of Jones's desire. If we are to understand Smith's and Jones's wants, we must be able to know (as surely we are) what has to be true for Smith's want to count as fulfilled, and similarly with Jones. If, as Geach claims, there is no proposition having a definite truth value available to us here as an expression of the objects of their wants, it is hard to see how this condition can be met.

This brings me to my second question concerning this part of Professor Geach's paper: how does he understand the connection which he claims to exist between the matter I have just been discussing and the truth of psychological egoism? He says: 'To enjoy . . . is to enjoy F-ing done by *oneself*: can one desire as an end what one could not enjoy?' Now it is not clear to me precisely how Geach thinks this important question is related to the issue hitherto discussed. Prima facie, it involves rather different considerations concerning the relation between the verbs 'to want' and 'to enjoy'.

Can one desire as an end what one could not enjoy? Even if one denies this, it will not follow that to desire something is to desire to enjoy it. Suppose for instance that one desires a benefit for some other person. Then if one comes to know that he obtains the benefit in question it may be assumed that one would 'enjoy' the knowledge of it. But from this it does not follow that it was this knowledge that one desired, as has been thought by some philosophers. Dr. Henry has made the point very elegantly and I should like to translate her remarks:

We are glad that the other person is happy and suffer from his being unhappy. Of course we must know of his happiness or unhappiness in order to feel with him. But . . . . it is not a matter of indifference to us that the other person should suffer, just as long as we do not hear about it. Thus our knowledge in this case is only the cause of our sympathetic feeling. But this feeling itself is directed towards what the other person is undergoing.[1]

Now some of the points of Geach's that I have briefly discussed *might* indeed be thought of as connected with this issue: but only by way of a confusion arising out of an ambiguity in the notion of 'satisfaction'. If a person desires something, then there is something that would count as a satisfaction of his desire. But *this* notion of 'satisfaction' has nothing directly to do with the psychological notion of 'enjoyment'. It is, let us say, a purely logical notion, pretty well identical with that involved in saying, for instance, of a man that he satisfies a certain description. It is, of course, true that, *in very many cases*, if someone's desire is, in this logical sense, satisfied, he will also gain psychological satisfaction: he will 'enjoy' the object of his desire. But the question whether this is so or not is nevertheless a different question from that which asks whether or not his desire has been satisfied or fulfilled; and psychological satisfaction *need* not accompany the satisfaction of one's desires.

Although this last issue is not one that baulks very large in Geach's paper, I have thought it worth dwelling on for a moment because of its very great importance in the philosophy of mind and in ethics. But I now turn to an issue more centrally embedded in Geach's overall argument. One of his main aims is to elucidate a sense of teleological explanation lacking the element of intentionality which manifests itself in the dependence of the truth value of a proposition on the 'mode of presentation' of its terms. Geach sees the importance of mode of presentation as primary when we are dealing with human thought and judgement, as less with wants, intentions,

[1] Grete Henry-Hermann: 'Die Überwindung des Zufalls' in Minna Specht and Willi Eichler (eds.), *Leonard Nelson zum Gedächtnis*, Verlag 'Öffentliches Leben', Frankfurt a.M.–Göttingen, 1953, p. 49.

and desires, and as altogether absent with 'pure teleology'. I shall consider first what he says about intentions and desires.

There is of course a distinction to be made between what an agent intends or desires and what he actually achieves. Geach does not of course deny this distinction, but his *argument* for the reduced importance of intentionality in connection with intentions and desires does seem to rely mainly on points he wishes to make about what is involved in an agent's achievement of a state of affairs. Let us consider these points. He says: 'the achievement of a state of affairs is independent of the mode of presentation: bringing it about that General de Gaulle is dead is the same thing as bringing it about that the Frenchman with the largest nose is dead if General de Gaulle is the Frenchman with the largest nose'. I already have qualms at this point. Is what I bring about by my action the same as what I 'achieve'? Consider the following case. A bomber crew returns from a mission and is told by the debriefing officer that through an error of navigation the ammunition dump which they have destroyed in fact belongs to their own side. That is what they have brought about by their action. The debriefing officer might say to them: 'What you have achieved by your action is the destruction of our most important reserve of ammunition.' There is nothing at all off-colour about this way of expressing the situation, but the crew would undoubtedly take is as bitterly ironical; and I think the dimension of irony is important here. It depends for its appropriateness on the fact that the crew may be presumed to have certain other ends: for instance, the harming of the enemy rather than of their own side. And the mode of presentation of those ends is of course vital to their *being* the crew's ends. The irony of the debriefing officer's use of 'achieve' derives precisely from the fact that what the crew brought about by their action was something they can have been presumed to have wanted very much not to happen. Their actions of course brought about many other states of affairs not obviously describable in terms which would relate them to the crew's wants and desires at all and concerning which, correspondingly, the use of the word 'achieve' would be quite out of place. The release of the bombs brought about a certain displacement of the air between the aircraft and the ground; it would not seem natural to me to say that this was something that the bomb-aimer 'achieved'; neither should I want to say that he pressed the release button 'in order that' the air should be displaced, even though it is certainly something that he 'brought about' by pressing the button.

What an agent brings about by his action cannot then in all cases be identified with what he achieves; and of course neither of these can be identified with what he intends or desires. How important is the mode of presentation of the object in these latter cases? I think we may be pulled in different directions according to the kind of example we make central and the point of view from which we consider it. Take Oedipus's desire (or intention) to marry Jocasta. Jocasta happens to be Oedipus's mother. So does Oedipus intend to marry his mother? Some considerations incline one to answer yes. Thus—the fact that Geach wishes to emphasize—if Oedipus does marry Jocasta, he marries his mother. And certainly that was his action, for which he bears responsibility: though it is not the same action or the same responsibility as would have been involved had he *known* that Jocasta was his mother. It is also important to note that the existence and continuation of Oedipus' intention depends on his *not* knowing that Jocasta is his mother. In other words the very existence of his intention depends on the mode in which Jocasta is presented to him. Desires are of course also subject to modification with changes in mode of presentation of their objects—though not necessarily to the same changes as are intentions.

Examples of different sorts could be multiplied concerning which we might want to say different things in respect of the role of mode of presentation to intention. I shall not try to do this here. Any attempt to sum up what is obviously a very complex situation in a general formula is bound to be hazardous, but my present inclination is to look at it in a way which is opposed to Geach's. I feel inclined to say that teleology (action for the sake of a certain end) comes in with the agent's own view of the situation and to that extent depends on the mode in which the object is presented to him. A third party may still speak teleologically in language which presents the object in a mode that is not the agent's, nor even perhaps available to him; but such a way of speaking has to be used with caution and may have to be suspended altogether in cases where the spectator's description of the situation presents the object in a mode which, had it been available to the agent, would have induced him radically to change his intention.

It is not altogether clear to me exactly how far this (admittedly rough and ready) formulation is in opposition to Geach's view. Geach does indeed acknowledge a 'tinge of intentionality' in such situations, but it looks as though, for him, its presence is not central to our understanding of the situations in teleological terms: such an understanding rests on our perception of the situation as it is and pretty well independently of the way it appears to the agent. (I have to use the evasive expression 'pretty well' here to register my unclarity about the extent of the role which Geach wishes to ascribe to the acknowledged 'tinge of intentionality'.)

It should be noted that the intentional opacity which does, according to my contention, belong to talk about the ends which agents seek by their actions does not conflict with my earlier insistence that the object of desires and intentions must be some determinate state of affairs specifiable in a proposition. The point is that in such a proposition the mode of presentation of its terms is relevant to its adequacy as an expression of the desire or intention in question.

The playing down of intentionality in the teleology involved in the actions of conscious agents is an important step in Geach's transition to 'pure' teleology, where he sees intentionality as vanishing altogether: that is to say, he thinks that here the mode of presentation of the end to which something occurred is entirely irrelevant. An important part of his case for saying this

is his account of propositions of the form '*a* brings it about that
p' which, on the one hand, are not truth-functional in that they
are true for some true interpretations of *p* and false for others,
but which are, on the other hand, extensional in the sense that
they have a 'direct relation to the way things are and the way
things act, rather than to aspects under which our minds
consider things'. He claims that the role of '*q*' in '*p* in order
that *q*' is connected with its role in '*a* brings it about that *p*' in
a way which makes the former propositions also, though not
truth-functional, nevertheless limitedly extensional. Now one
can allow the connection between the roles of '*q*' in these two
sorts of case and still wonder whether the connection really
produces the result which Geach claims. Suppose we say:
'This trigger mechanism is so arranged in order that the cartridge may explode when the trigger is pressed.' If the weapon
is faulty in some way, the explosion of the cartridge may also
'present itself' as an explosion of the whole weapon resulting
in injury to its user. Here I should certainly say that the trigger
mechanism had been so constituted as to bring about both the
explosion of the cartridge and also the explosion of the weapon.
But though I should say that it has been so constituted 'in
order that' the first result should occur, I should not feel inclined to say the same of its relation to the second result
unless I had reason to suspect, for example, sabotage, that is,
according to my previous argument, something intentional.

In developing his argument on this point Geach makes a
criticism of Mill which seems to me partly right and partly
wrong. He rejects Mill's claim that teleological notions are
applicable to, say, a clock only because we have abundant inductive evidence that pre-existing human desires and intentions
are a condition of the existence of such things. Geach seems to
me quite right, and to make an important point, when he remarks that 'of course nobody has collected such evidence—
we just feel sure that we could collect it if we wished. No doubt
we could, but our insight into the teleology of the clock does
not wait upon this and we might have the same insight if

unexpectedly we picked up such an apparatus of clearly non-human origin on the sands of Mars'. But he also seems to me to attach insufficient importance to our confidence that evidence of this sort *could* be collected. I feel inclined to say that to the precise extent to which such confidence comes to be shaken by further investigation, so also will our inclination to describe the situation teleologically be shaken. I do not want to deny, or deny the importance of, the fact that in some circumstances we may feel unshakably certain that something before us is the product of intentional intelligent agency, even where we find ourselves completely unable to find any external supporting evidence for such a belief. Nor do I want to deny that this certainty may be based on an 'insight into the teleology of the clock' which we have arrived at simply by close investigation of the clock itself. Our *evidence* for this teleology lies in the clock itself; but what makes what we see in the clock into such evidence is the background of knowledge and presumptions against which we investigate the clock. This background is at once brought into play by our observation of well-tooled metallic parts, precisely fitted together metallic shapes producing systematically interrelated motions, signs of lubrication and the like: things which we just do not meet in nature. So while there is a strong case for saying that our application of a teleological schema here is based on evidence internal to the object before us, this should not close our eyes to the facts in the background which lie behind our treating what we see as evidence for this sort of conclusion. And, as I have suggested, were we to find external evidence strongly suggesting that we are wrong in placing what we see against a background of intelligent intentional action, our initial interpretation of what we see would be called in question. This point is not affected by the fact that, with things as they are, we may feel completely confident that nothing of the sort will be found, or that, if it is, some other evidence will be found which overrides it.

Geach claims that teleological ways of thinking are indispensable, first in our understanding of living things, and second in our understanding of natural processes generally. He does not devote a great deal of space to the question of teleology in biology and it is not altogether clear to me how far he regards this as a special case, requiring different treatment from, say, physics. The issues here are of course complex and difficult, involving amongst other things a treatment of the proper application and interpretation of so-called 'functional' explanations. I shall follow Geach in not discussing these issues and content myself with remarking that he seems to me perfectly right in insisting on the naturalness with which we find ourselves thinking in a teleological way in respect of at least certain kinds of process in living organisms, as also in the heuristic value of such ways of thinking. But his more interesting and radical claim is that teleological notions are needed to make intelligible our thinking in terms of *efficient* causation, if I understand him correctly, right across the board in the natural sciences. His case for this rests on the interesting use to which he puts Kenny's account of practical reasoning, from which he singles out the point that the conclusion of a chain of practical reasoning, unlike that of a purely 'theoretical' deduction (as in geometry), is 'defeasible' by the addition of extra premises. Geach makes the extremely important point that something similar is true in connection with explanations in terms of efficient causation, our understanding of which involves the realization that the operation of a cause may be counteracted or interfered with by some further intervening cause. He seems to me perfectly right in insisting on the difficulties this fact creates for the hypothetico-deductive interpretation of natural science. I remain unconvinced, however, that this is enough to justify us in saying that any 'teleology' is involved here.

Geach puts his point in terms of the notion of a 'tendency'. It is because explanations in terms of efficient causation are defeasible that '(w)hat laws tells us is what tendencies will exist when certain agencies operate'; and tendency, he claims, is a teleological notion. He connects his position with Aquinas's remark that 'every agent acts for an end', which he re-expresses in the form: 'There is no coherent way of describing behaviour unless we describe it as acting in order that so and so may happen.' His claim about the teleological significance of 'tendency' seems to lead him to identify this last phrase with the expression: 'tending to act in a certain way'.

Of course there is no doubt that *some* tendencies are teleological, for instance the tendency of a thermostatic mechanism to maintain an even temperature. But *all* tendencies? Bodies within the earth's gravitational field tend to move towards the centre of the earth unless prevented. Do they so move *in order that* they may arrive at the centre of the earth? It would not seem to me natural to say so; nor can I see that our understanding of the efficient causality involved in such motion is increased by our saying so. Of course, if all that is intended is to point to the defeasibility involved in our reasoning about the operation of such causality, that can be accepted. But then it seems to me misleading to express this point in terms of any 'teleology'. We might remind ourselves here of the important point Geach himself makes about the ambiguity of talk about the 'end' of a process as signifying either the final stage of the process or that for the sake of which the process occurs. What Geach points out is that the end of a process in this second sense does not necessarily have to be its end in the first sense. But this cuts both ways: the final stage of a process, or what happens after, as a result of, a process does not have to be something for the sake of which the process occurs.

Comment

BY GRETE HENRY

Professor Geach is perfectly right when pointing out that we cannot avoid teleological thinking and that it is therefore essential that we should engage in it consciously and critically in order not to be led astray. I should like to comment on this idea by advancing a few doubts and a number of supplementary suggestions.

In his attempt to prove that we cannot avoid forming teleological concepts the speaker stressed the fact that even in theoretical explanations in which events are explained in terms of causality we are unable to dispense with such concepts. The following of effects upon causes, so he maintained, is by no means an invariable succession of events; a law of nature does not state what *de facto* always happens. What he means by this is presumably: what happens *under given conditions* which, according to this law, have certain effects. A law of nature, he said, states what happens under such conditions unless those effects are annulled or altered by prevention or interference. Thus, according to Professor Geach, the insight into the law of nature merely represents an insight into a tendency, that is a tendency for those events to happen—under the conditions mentioned—unless prevented or interfered with. The notion of a tendency, however, is a teleological notion.

Such is the gist of Professor Geach's argumentation. Now the first objection I should like to raise concerns the distinction he draws between on the one hand prevention and interference modifying the events in nature and on the other hand the conditions causing the events according to the law of nature. Where are we to draw the line? Let us consider, by way of example, the law of gravity in classical mechanics and its formula $s = \frac{1}{2} g.t^2$ which says that in free fall all bodies fall at the same acceleration; thus in the first second of their fall they all cover roughly 5 metres. Now, this happens only 'if nothing else interferes', otherwise it is not a 'free' fall. However, in everyday life something does interfere in the majority of instances of a body falling, be it only the air in which we live and in which stones fall faster than feathers. The only way in which the free fall can be demonstrated to students in a physics lesson is in the artificially produced conditions of a vacuum, as for instance the fall of a stone and a feather in the 'closed system' (that is, sufficiently 'closed' for this experiment) of a glass pipe pumped empty of air. Students tend to be fascinated by the fact that in this pipe the stone and the feather fall with equal speed. But they certainly do not see that the earth has always had the tendency to attract falling bodies with equal acceleration and that it is merely prevented from doing so because it is surrounded by air. The difference between the fall in the glass pipe emptied of air and that in the air of the class-room is explained theoretically by reference to the different starting conditions, the different *causae efficientes*, while the only teleological 'tendency' or *causa finalis* that might be mentioned in the total context is the teacher's intention to demonstrate the free and the impeded fall. This intention determines his physics teaching but not the lawlike nature of the processes of fall, and only the latter are the subject-matter of his teaching.

Thus I cannot accept the speaker's claim made at the end of his lecture that even 'efficient causality' which is determined by laws of nature cannot be adequately described without teleological notions, although I do realize that he was not talking about the laws of classical mechanics—as I have been doing now—but about laws of biology: he showed that the circulation of the blood is explained far better by Harvey and his teleological interpretation than by Descartes who methodologically excluded all teleological explanations from natural science. Professor Geach also pointed out that people who explain the development of living beings according to Darwin in terms of mutation and selection and who claim that this makes all teleological interpretations superfluous and misleading, are themselves reintroducing teleological terminology when saying that 'nature' or 'evolution' pursues this or that 'purpose' and opens up new ways and means of solving the problems of life. Such personifications are evidently instances of people being led astray in the sense referred to by the speaker and which we are to avoid by means of a conscious and critical use of teleological notions. Thus he offers the logico-grammatical form '*p* in order that *q*' as the form of the teleological judgement by means of which life processes can be described and explained.

Now, the first difficulty I should like to bring to your attention is that important spheres of biological research do not in fact lend themselves to such treatment and that scientists are looking for theoretical explanations of processes in the living organism; this applies to organic chemistry as well as to the young science of molecular biology. A more serious difficulty seems to me to arise from the fact that the teleological form '*p* in order that *q*' is from the very outset treated without consideration for any agent who might bring about *p* in order that *q* should happen. We shall have to ask ourselves whether and how this interpretation is compatible with the speaker's initial stipulation that teleological explanations are explanations in terms of ends, and ends are related to things we want. It seems to me that by detaching the notion of teleological connection from that of a subject's purposeful behaviour he blurs the difference brought out by the distinction between efficient and final causation, between *causa efficiens* and *causa finalis*.

Let us therefore consider the different connections which we have in mind when speaking of the one and the other form of 'causation'. This contrast to my mind comes out most clearly when we consider the difference between two modes of making fairly reliable predictions concerning future events. Both can be observed in nearly all prudent behaviour, but as a rule they are so closely connected that it is hard to draw a clear line between them. I shall therefore illustrate this difference by two seemingly primitive examples: on the one hand our knowledge of certain data and laws of nature allows us to predict that in the next weeks and months till 21 December the sun will rise progressively later in the northern hemisphere; on the other hand familiarity with final causation enables me to predict that in a moment, after this announcement, I shall count to three and clap my hands. (Doing it.) In the second case, the prediction was successful because I actually performed a connection of the form '*p* in order that *q*'. Here *p* stands for the announcement and performance of this little demonstration, *q* for having at my disposal a sufficiently simple and unambiguous example of a prediction, based on a teleological connection '*p* in order that *q*', which is shown to be successful and was not deduced by causal considerations.

I am again led to think of this distinction between two modes of predicting future events by the speaker's contrasting theoretical and practical reasoning in his attempt further to elucidate the concept of teleological explanation. He suggests that in contrast to theoretical reasoning practical reasoning is 'defeasible'.

What does he mean by that? As he denies 'defeasibility' to theoretical reasoning he cannot mean adducing of proof that errors or logical mistakes have been made, but merely the invalidation of correctly performed reasoning based on valid premises. Nor can he mean that our expectations and predictions change or prove to be mistaken by additional experience, even where the former are not at all blind but well-considered—as is frequently the case with weather predictions based on careful metereological considerations. Theoretical arguments can only be denied 'defeasibility' in the sense of 'refutability' if all that is to count as such arguments are cogent conclusions from well-established premises. Only then is it true—as the speaker argues it is—that in the sphere of theoretical reasoning no added, newly acquired insight and experience can cancel out earlier findings.

One thing I want to mention in passing: in the face of the numerous revisions of scientific theses that have taken place in this century, particularly in physics, the question arises whether and in what sense theoretical reasoning and its scientifically formulated findings really do lack the characteristic of 'defeasibility' and thus have the characteristic of well-established finality. But this is a wide field and takes us beyond the scope of our present argument.

It remains true, however, that only in the sphere of practical reasoning can the addition of a new insight or experience be sufficient to overthrow an otherwise well-founded intention. The teleological connection 'p in order that q' can be invalidated by the statement that p without p' cannot be had and that p' would be too high a price to pay for q. The teleological connection 'p' in order that q' can be defeated and rejected on the strength of the counter-decision 'non-q in order that non-p''. Nothing analogous occurs in logically sound and carefully formulated theoretical arguments.

Any attempt to deal with teleological explanations in a scientific and systematic manner will have to take this feature into consideration. For in cases where there is an alternative—whether to put up with or even bring about p for the sake of q or to do without q in order to avoid p—the question is which it would be better to do. This is a question of value and it concerns human behaviour. It is therefore in the last resort —marginally or centrally—a question belonging to the realm of ethics, insofar as we are considering the objective decidability of the problem.

Thus teleology as a science exists if ethics can be developed as the science of the value and non-value of human behaviour. Within the School of Critical Philosophy which based itself on Kant my teachers Jakob Friedrich Fries and Leonard Nelson were led by this to characterize and systematically develop ethics as 'subjective teleology', that is teleology which is related to the actions and behaviour of a subject in search of value criteria that will allow his autonomy in the way he lives his life. In Critical Philosophy this 'subjective teleology' is contrasted with 'objective teleology', namely the wide range of problems concerning the value and meaning of happenings in general. However, Critical Philosophy merely shows that such problems defy any scientific and systematic treatment in terms of clearly defined concepts: they can only find their answers in religious trust. I can merely hint at this, as any further elaboration would take me too far away from our present concern.

In concluding I should like to come back once again to that

sphere of teleological statements, above all in biology, where the basic form 'p in order that q' is treated without any reference to an agent who realizes or permits p in order that q. Wherever processes in a living organism are understood as functions fulfilled by its organs in order that the living being itself or its species should be preserved, we are clearly dealing with a different kind of connection from that in cases where the connecting 'in order that' is related to an agent who, in a situation of conflict, is confronted with the necessity to make a decision. It therefore seems that there are two different forms of the teleological connection 'p in order that q'. They are so closely related that we can meaningfully apply the same linguistic representation; yet they are so different from each other that when transferring the notion of the purposeful action of an agent to our teleological interpretation of life processes—a step suggested by our linguistic usage—we may be led to indulge in those deceptive personifications according to which Nature has purposes and pursues them.

What, then, is the relation between these two teleological notions? Well, after all it is living beings who perform 'p in order that q' in purposeful actions, and they are able to do so because they have at their disposal and employ certain physical and mental powers. Although crafts and technology have long extended the individual's range of action beyond that which is accessible without tools and devices intentionally created by man, the tools created 'in order that' they may be purposefully employed do not replace man's limbs and senses but merely support them. Vice versa, we can see the limbs and senses of our organism in analogy to tools and thus interpret their functions teleologically: we have eyes in order to see, hands in order to grasp. And the next step in the transferring of the purposeful character of our own behaviour on to the conditions determining it takes us still further away from our consciously performed actions: the circulation of the blood is there in order to feed the cells of our body, and the heart is there in order to keep up the circulation of the blood.

The fact that we borrow teleological ideas from the consciousness of our own purposeful actions and transfer them on to other processes is no argument against their being used in this way. Here I agree with Professor Geach that the use of such teleological notions is necessary for our understanding of life processes and living organisms and is perfectly legitimate. This does not mean, however, that the same life processes should not also be looked into with the intention of providing a theoretical explanation. The inquiry into chemical and physical connections in which life processes take place according to laws of nature is a scientifically meaningful, indeed a necessary task. Neither form of research excludes the other, and it is our task to relate them to each other as well as to distinguish between them. Again I am unable to do more than hint at this point, but we might have occasion to pursue it in our discussion.

Reply to Comments

BY PETER GEACH

I hoped to have made it clear that the main aim of my paper was to defend the view that teleological patterns of explanation apply to inanimate and unconscious natural processes. Professor Winch clearly has other interests, and little sympathy with this aim. A number of points that he raises are inherently interesting, but little relevant to my original topic; I must excuse myself from chasing after these new hares he has started, in a reply that must be reasonably brief.

That an end is an object of desire and intention is a grammatical remark, in Wittgenstein's sense: this is what it *makes sense* to say is desired or intended—even if the ends of nature, which were my concern, are not actually the objects of desire or intention, even then the description of them must have this logical grammar. It would have been simple to say that the end is always linguistically expressed in a proposition; but I do not wish to sacrifice truth for simplicity's sake, and for this reason I mentioned the complication of self-regarding desires and intentions, and the logical problems they raise. For ends in nature, my main concern, this complication does not arise; so it is a pity that Winch has devoted so much space to it.

I am afraid Winch has not grasped the logical point I was making about the matter. Consider the proposition: 'Anybody who desires that he himself should marry Miss Brown is very silly.' It is of course 'impossible to specify what state of affairs would count as the satisfaction of the desire in question', as Winch protests: such a specification would require saying *whose* marriage to Miss Brown we were talking about, and in my example there can be no answer to that question. All the same the proposition is quite intelligible. Obviously its analysis raises logical problems; these have to be faced, not dismissed. I do indeed 'claim' that the content of the *that* clause in this proposition cannot be given by replacing it with (the appropriate grammatical form of) a proposition with a truth-value; I still think so, and I deny the need to meet the condition Winch says must be met.

My question about psychological egoism came in a parenthesis; it interests Winch, and his answer would deserve discussion; but the answer has nothing to do with ends in nature, so I shall not go further into the matter.

I had reason to say that the mode in which a situation presents itself to a subject is *less* significant as regards desire and intention than as regards belief: the fulfilment-conditions of a desire or intention do more to determine *what* is desired or intended, the content of the desire or intention, than the truth-conditions of a belief do to determine *what* is believed. This is obvious as regards the desires and intentions we ascribe to brutes: if a stallion wants to cover the mare in the next field, and the mare in the next field is in fact his mother, then he wants to cover his own mother, and to ask whether she is presented to him *as* the mare in the next field or *as* his mother would be silly. Even as regards human intentions, the way the situation presents itself to the agent is not *decisive* about what he wants or intends; and self-deception about what you want or intend is accordingly much easier to understand than self-deception that you believe that *p*, when in fact you do not believe that *p*. To persuade yourself that you believe *p*, when you do not in fact believe *p*, means getting into a very odd state of mind: to deceive yourself about your intentions, you need only go in for mental exercises of 'directing the intention', as described in Pascal's *Provincial Letters* or listed in papal condemnations contemporary with Pascal.

I am not sure how Winch intends his distinction between bringing about and achieving. In actual usage the two verbs of course overlap: a stipulative distinction might all the same be useful for philosophical discussion—but Winch has not given any stipulation, only examples, and I do not know how to go on from these. The distinction, whatever it may be, could hardly make sense for the inanimate and unconscious agents that were my main concern.

Even as regards these, Winch confusingly brings in human attitudes. Consider his example 'This trigger mechanism is so arranged that the cartridge may explode when the trigger is pressed'—'. . . that the whole weapon may explode'. In the set-up Winch is envisaging, there is nothing to choose between the two ascriptions of teleology, so long as we firmly concentrate on the tendencies of the inanimate agents involved. Of course the purpose and interests of men may enter differently: the explosion of the cartridge may be designed and the explosion of the whole weapon accidental. I do not deny this difference, but for my subject of discussion it is not relevant; the cartridge and the gun know nothing of men's purposes.

The example 'Bodies within the Earth's gravitational field move in order that they may arrive at the centre of the Earth' is one that Winch has wished upon me, not one that I gave; he reminds me of my own point that the final stage of a process need not be what it is for, but this is a reminder anyhow out of place here, since falling bodies do not end up at the Earth's centre.

The one example of inanimate teleology that Winch does take from me is the clock on the sands of Mars. The efficacy of reasoning like this:

Let's suppose this is a mechanism to make the hands revolve at uniform rates. If so, there would have to be a mechanism to correct for temperature changes—and there *is* a mechanism that would do just that. Etc., etc.

is to be judged by our success in describing the actual structure of the clock, and is not then refutable by failure to find any further evidence of Martian clock-makers. I should say the same, *mutatis mutandis*, about Harvey's successful teleological account of the circulation. But Winch does not follow me in discussing this.

Professor Henry seems not to have understood my use of the term 'defeasible'. This is an English lawyer's term of art; and perhaps it has become something of a vogue word, used carelessly and inexactly. But my own use of it was close to the lawyer's use. A marriage or a contract may stand as valid if unchallenged, but be defeasible or voidable if some fault in it is established in court. I maintained that in an analogous sense practical reasoning, though not theoretical deductive reasoning, is defeasible: practical reasoning from a set of directives as premises is defeasible by the addition of a premise if its conclusion is incompatible with the fulfilment of that premise, but stands firm if no such premise is added; whereas theoretical reasoning is *never* defeasible by the addition of a premise. That, at least, was my contention. Professor Henry's reference to the revision of scientific theories is not to the point; in such a case the premises are not added to but altered. Even her account of how practical reasoning can be defeated was not in line with my thought. For I was considering an idealized case of practical reasoning in which all the directives used as premises can consistently be fulfilled, even when the premises are augmented; in this model, all the premises are logically on a level, none more equal than any other; so the question of greater and less value of ends, raised by Professor Henry, just does not come in.

Causal reasoning, I maintained, is defeasible in a way closely analogous to practical reasoning. Professor Henry asks where we are to draw the line 'between on the one hand prevention and interference modifying the events in nature and on the other hand the conditions causing the events according to the laws of nature'. I do not think of drawing a line *here*; a causal factor in causal reasoning, like a directive in my idealized practical reasoning, is a premise on the same level as other premises. The line is drawn in deciding not to bring in other factors that might affect the result; and this line has to be drawn somewhere by a human reasoner, at the risk that the excluded factors may defeat the conclusion.

From this defeasibility it follows that the formal structure of causal reasoning is after all parallel to that of practical reasoning. We could ascribe to nature certain ends to be fulfilled, corresponding to the different causal factors, and derive what actually happens as a way of reconciling and fulfilling these diverse ends. This may appear artificial; but in scientific practice, reasoning from such principles as le Chatelier's, or again as the principle of least action, actually *looks* teleological—a fact that gave needless scandal to Poincaré in *La science et l'hypothèse*. And the failure to realize this defeasible structure of causal reasoning leads to errors that are not harmless. Physics cannot in principle be stated *more geometrico*; for geometrical reasoning is in principle indefeasible—what really does follow from limited information about the attributes of a figure absolutely and is not defeasible by further information. Contrary to what Professor Henry supposes me to think, I do defend the Aristotelian teleology in physics as well as in biology—not that Aristotle was not very fallible about *which* teleologies actually hold good.

A structure of inference *formally* parallel to the practical reasoning of a man pursuing various (reconcilable) goals may serve our purpose in describing the order of nature without our having to assume that either natural unconscious agents, or their Maker, must deliberate how to adapt means to ends. The logic of practical inference can be as well detached from the psychology of desire as the logic of theoretical inference from the psychology of belief. When we make such constructions, we may be tempted to understand them animistically, or again in terms of Paley's natural theology; but the temptation, once we are conscious of it, can be resisted, and our yielding to it is no condition for the usefulness and heuristic value of teleological schemata. On this last point, I hope I am right in finding Professor Henry in agreement with me.

III/Theoretical Explanation

Wesley C. Salmon[1]

In previous discussions of the explanation of particular events,[2] I have argued—contra Hempel and many others—that such an explanation is not 'an *argument* to the effect that the event to be explained . . . *was to be expected* by reason of certain explanatory facts' (my italics).[3] Indeed, in the case of 'inductive' or 'statistical' explanation at least, I have maintained that such explanations are not *arguments* of any kind, and that consequently, they need not embody the *high* probabilities that would be required to provide reasonable grounds for expectation of the explanandum event. I have argued, instead, that a statistical explanation of a particular event consists of an assemblage of factors relevant to the occurence or non-occurrence of the event to be explained, along with the associated probability values. If

[1] The author wishes to express his gratitude to the National Science Foundation (U.S.A.) for support of research on scientific explanation and other related topics.
[2] 'Statistical Explanation' in *The Nature and Function of Scientific Theories*, Robert G. Colodny, ed. (Pittsburgh: University of Pittsburgh Press, 1970), and reprinted in Wesley C. Salmon, et al., *Statistical Explanation and Statistical Relevance* (Pittsburgh: University of Pittsburgh Press, 1971).
[3] Carl G. Hempel, 'Explanation in Science and in History', in *Frontiers of Science and Philosophy*, Robert G. Colodny, ed. (Pittsburgh: University of Pittsburgh Press, 1962), p. 10.

the probabilities are high, as they will surely be in some cases, the explanation may provide the materials from which an argument can be constructed, but the argument itself is *not* an integral part of the explanation. This model has been called the 'statistical-relevance' or '*S–R* model'.[4]

In addition, I have claimed that the so-called 'deductive-nomological' model of explanation of particular events is incorrect. It is not merely that there are explanandum events which seem explainable only inductively or statistically; Hempel and Oppenheim acknowledged such cases from the very beginning. There are also cases—such as the man who consumes his wife's birth control pills and avoids pregnancy—in which an obviously defective explanation fulfils the conditions for deductive-nomological explanation. All such examples seem to me to exhibit failures of relevance. I have suggested, therefore, that even events which appear amenable to deductive-nomological explanation should also be incorporated, as limiting cases, under the statistical-relevance model.[5]

Arguments by Greeno and others[6] have convinced me that explanations of particular events seldom, if ever, have genuine scientific import (as opposed to practical value), and that explanations which are scientifically interesting are almost always explanations of classes of events. This leads to the suggestion, elegantly elaborated by Greeno,[7] that the goodness or utility of a scientific explanation should be assessed with respect to its ability to account for entire classes of phenomena, rather than by its ability to deal with any particular event in

[4] See especially the Introduction of *Statistical Explanation and Statistical Relevance*, op. cit.

[5] This approach has been elaborated in some detail in the three essays by Richard C. Jeffrey, James G. Greeno, and myself in *Statistical Explanation and Statistical Relevance*.

[6] For example William P. Alston, 'The Place of Explanation of Particular Facts in Science', *Philosophy of Science*, XXXVIII, 1 (March 1971), pp. 13–34.

[7] James G. Greeno, 'Explanation and Information', in *Statistical Explanation and Statistical Relevance*; first published as 'Evaluation of Statistical Hypotheses Using Information Transmitted', *Philosophy of Science*, XXXVII, 2 (June 1970), pp. 279–93.

isolation. If, to use Greeno's example, a sociological explanation is offered to account for delinquent behaviour in teen-age boys, it is to be evaluated in terms of its ability to assign correct probability values to this occurrence among various specifiable classes of boys, not in terms of its ability to predict whether Johnny Jones will turn delinquent. This shift of emphasis is important, because it removes any temptation to suppose that we cannot explain Johnny's behaviour unless we can cite conditions in relation to which it is highly probable. Perhaps Johnny is a member of a class in which delinquency is very improbable, and no more can be said in the matter. This does not mean that the explanation of *his* delinquency—which is just part of the explanation of delinquency in boys—is defective or weak. As Jeffrey has argued persuasively,[8] the explanation of a low-probability event is not necessarily any weaker than the explanation of a high-probability event. Even if Billy Smith is a member of a class of boys in which the delinquency rate is very high, the explanation of his delinquency by the above-mentioned sociological theory is no better or stronger than the explanation of Johnny Jones's delinquency. High probability is not the desideratum, nor is it the standard by which the quality of explanations is to be judged; rather, a correct probability distribution across *relevant* variables is what we should seek.

At the conclusion of my elaboration of the *S–R* model, I expressed certain reservations about it. The two most important problems concerned the involvement of causality in scientific explanation and the nature of theoretical explanation. These two problems are intimately related to one another, and together they form the subject of the present paper. I shall agree from the outset that *causal relevance* (or causal influence) plays an indispensable rôle in scientific explanation, and I shall attempt to show how this relation can be explicated in terms of the

---

[8] Richard C. Jeffrey, 'Statistical Explanation vs. Statistical Inference', in *Statistical Explanation and Statistical Relevance*; first published in *Essays in Honour of Carl G. Hempel*, Nicholas Rescher, ed. (Dordrecht, Holland: D. Reidel Publishing Company, 1969), pp. 104–13.

concept of statistical relevance. I shall then argue that the demand for suitable causal relations necessitates reference to theoretical entities, and thus leads to the introduction of theoretical explanations. The theme of the paper will be the centrality of certain kinds of *statistical* relevance relations in the notions of casual explanation and theoretical explanation. The result will be an account of theoretical explanation that differs fundamentally from the received deductive-nomological model.[9]

I THE COMMON CAUSE PRINCIPLE

When all of the lights in a room go off simultaneously, especially if quite a number were on, we infer that a switch has been opened, a fuse has blown, a power line is down, or so forth, but not that all of the bulbs burned out at once. It is, of course, possible that such a chance coincidence might occur, but so improbable that it is not seriously entertained. The principle is not very different from that by which we conclude that two (or five thousand) identical copies of the same book were produced by a common source. A similar kind of inference is involved when one observes an ordinary bridge deck arranged in perfect order, starting from the ace of spades, and concludes (knowing that cards are packed that way at the factory) that this is a newly opened unshuffled deck, rather than one which arrived at the orderly state by random shuffling. The same principle is involved when two witnesses in court give testimony which is alike in content; if collusion can be ruled out, we have strong grounds for supposing that they are truthfully reporting something they both have observed.

The principle governing these examples has been pointed out by many authors. It is deeply embedded in Russell's infamous

[9] Richard Beven Braithwaite, *Scientific Explanation* (New York and London: Cambridge University Press, 1953); Carl G. Hempel, *Aspects of Scientific Explanation* (New York: The Free Press, 1965); Ernest Nagel, *The Structure of Science* (New York: Harcourt, Brace & World, 1959) are among the major proponents of the 'received view'.

'postulates of scientific inference',[10] and Reichenbach has called it 'the principle of the common cause'.[11] It may be stated roughly as follows: When apparently unconnected events occur in conjunction more frequently than would be expected if they were independent, then assume that there is a common cause. This principle demands considerable explication, for it involves such obscure concepts as *cause* and *connection.*

Let us take our departure from the standard definition of *statistical independence.* Given two types of events A and B that occur, respectively, with probabilities $P(A)$ and $P(B)$, they are statistically independent if and only if the probability of their joint occurrence, $P(A \& B)$, is simply the product of their individual occurrences; i.e.

$$P(A \& B) = P(A) \times P(B)$$

If, contrariwise, their joint occurrence is more probable (or less probable) than the product of the probabilities of their individual occurrences, we must say that they are not statistically independent of one another, but rather, that they are *statistically relevant* to each other. Statistical independence and statistical relevance, as just defined, are clearly symmetric relations.

It seems fairly clear that events which are statistically independent of each other are completely without explanatory value with regard to one another. If, for example, recovery from neurotic symptoms after psychotherapy occurs with a frequency equal to the spontaneous remission rate, then psychotherapy has no explanatory value concerning the curing of mental illness.[12] One reason that independence is of no help whatever in providing explanations is that independent

[10] Bertrand Russell, *Human Knowledge: Its Scope and Limits* (New York: Simon and Schuster, 1948), Part VI, esp. ch. IX.

[11] Hans Reichenbach, *The Direction of Time* (Berkeley and Los Angeles: University of California Press, 1956), §19.

[12] This thesis is argued at length in my essay, 'Statistical Explanation', op. cit.

events are inferentially and practically irrelevant; knowing that an event of one type has occurred is of no help in trying to predict the occurrence or non-occurrence of an event of the other type, or in determining the odds with which to bet on it. Another reason, which will demand close attention, is that statistically independent events are causally irrelevant as well.

If events of the two types are not independent of one another, the occurrence of an event of the one type *may* (but *need* not) help to explain an event of the other type. Suppose, for instance, that the picture on my television receiver occasionally breaks up into a sort of herringbone pattern. At first I may think that this is occurring randomly, but I then discover that there is a nearby police broadcasting station that goes on the air periodically. When I find a strong statistical correlation between the operation of the police transmitter and the breakup of the picture, I conclude that the police broadcast is part of the explanation of the television malfunction. Roughly speaking, the operation of the police transmitter is the cause (or a part of the cause) of the bad TV picture. Obviously, a great deal more has to be filled in to have anything like a complete explanation, but we have identified an important part.

In other cases, however, statistical correlations do not have any such direct explanatory import. The most famous example is the barometer. The rapid dropping of the barometer does not explain the subsequent storm (though, of course, it may enable us to predict it). Likewise, the subsequent storm does not explain the behaviour of the barometer. Both are explained by a common cause, namely, the meteorological conditions that cause the storm and are indicated by the barometer. In this case, there is a statistical relevance relation between the barometer reading and the storm, but neither event is invoked to explain the other. Instead, both are explained by a common cause.[13]

The foregoing two examples, of the TV interference and the

[13] The barometer example is analysed in some detail in *Statistical Explanation and Statistical Relevance*, pp. 53-5.

barometer, illustrate respectively cases in which correlated events can and cannot play an explanatory rôle. The difference is easy to see. The instance in which the event can play an explanatory rôle is one in which it is cause (or part thereof) of the explanandum event The case in which the event cannot play an explanatory rôle is one in which it is not any part of the cause of the explanandum event.

Reichenbach's *basic* principle of explanation seems to be this: *every relation of statistical relevance must be explained by relations of causal relevance.* The various possibilities can be illustrated by a single example. An instructor who receives identical essays from Adams and Brown, two different students in the same class, inevitably infers that something other than a fortuitous coincidence is responsible for their identity. Such an event might, of course, be due to sheer chance (as in the simultaneous burning out of all light bulbs in a room), but that hypothesis is so incredibly improbable that it is not seriously entertained. The instructor may seek evidence that one student copied from the other; i.e. that Adams copied from Brown or that Brown copied from Adams. In either of these cases the identity of the papers can be explained on grounds that one is cause (or part of a cause) of the other. In either of these cases there is a direct causal relation from the one paper to the other, so a causal connection is established. It may be, however, that each student copied from a common source, such as a paper in a fraternity file. In this case, neither of the students' papers is a causal antecedent of the other, but there is a coincidence that has to be explained. The explanation is found in the common cause, the paper in the file, that is a causal antecedent to each. (What university teacher has not witnessed the consternation of two students confronted with the identity of their papers when neither had copied from the other, but both had used a common file; this nicely dramatizes both the need for an explanation and the practical certainty that a common cause exists.)

The case of the common cause, according to Reichenbach's

analysis, exhibits an interesting formal property. It is an immediate consequence of our foregoing definition of statistical independence that event A is statistically relevant to event B if and only if $P(B) \neq P(A, B)$.[14] Let us assume positive statistical relevance; then

$$P(A, B) > P(B) \text{ and } P(B, A) > P(A).$$

From this it follows that

$$P(A \& B) > P(A) \times P(B).$$

To explain this improbable coincidence, we attempt to find a common cause C such that

$$P(C, A \& B) = P(C, A) \times P(C, B),$$

which is to say that, in the presence of the common cause C, A and B are once more rendered statistically independent of one another. The statistical dependency is, so to speak, swallowed up in the relation of causal relevance of C to A and C to B. Under these circumstances C must, of course, be statistically relevant to both A and B; that is,

$$P(C, A) > P(A) \text{ and } P(C, B) > P(B).$$

These *statistical* relevance relations must be explained in terms of two causal processes in which C is *causally* relevant to A and C is *causally* relevant to B.

A further indirect causal relation between two correlated events may obtain, namely, both may serve as partial causes for a common effect. Perhaps Adams and Brown are basketball stars on a championship team which can beat its chief rival if and only if either Adams or Brown plays. Caught at plagiarism, however, both are disqualified and the team loses. As Reichenbach points out, a common effect which follows a combination of partial causes cannot be used to explain the coincidence in the absence of a common cause. In the absence of any common

[14] As is my wont, I am following Reichenbach's non-standard notation, using '$P(A, B)$' to stand for the probability *from A to B*, i.e. the probability, given A, of B.

source, and in the absence of copying one from the other, we cannot attribute the identity of the two papers to a conspiracy of events to produce the team's defeat.[15] Thus, there is no 'principle of the common effect' to parallel the principle of the common cause. This fact provides a basic temporal asymmetry of explanation which is difficult to incorporate into the standard deductive-nomological account of explanation.

2 CAUSAL EXPLANATION OF STATISTICAL RELEVANCE

To provide an explanation of a particular event we may make reference to a statistically relevant event, but the statistical relevance relation itself is a statistical generalization. I agree with the standard nomological account of explanation which demands that an explanation have at least one general statement in the explanans. As indicated in the preceding section, however, we are adopting a principle which says that relations of statistical relevance must be explained in terms of relations of causal relevance. This brings us to the problem of explanations of general relations.

Most of the time (though I *am* prepared to admit exceptions) we do not try to explain statistical independencies or irrelevancies. If the incidence of sunny days in Bristol is independent of the occurrence of multiple human births in Patagonia, no explanation seems called for.[16] Statistical dependencies often do demand explanation, however, and causal relations constitute the explanatory device. Plagiarism, unfortunately, is not a unique occurrence; identical papers turn up with a frequency that cannot be attributed to chance. In such cases it is possible

---

[15] The temporal asymmetry of explanation is discussed at length in connection with the common cause principle (and lack of a parallel common effect principle) in 'Statistical Explanation', §11–12.

[16] In a situation in which we expect to find a statistical correlation and none is found, we may demand an explanation. Why, for example, is the presence of a certain insecticide irrelevant to the survival of a given species of insect? Because of an adaptation of the species? or an unnoticed difference between that species and another that finds the substance lethal? Etc.

to trace observable chains of events from the essays back to a *causal* antecedent. In these instances nothing of a theoretical nature has to be introduced, for the explanation can be given in terms of observable events and processes.[17] In other cases, such as the breakup of the television picture, it is necessary to invoke theoretical considerations if we want to give a causal explanation of the statistical dependency. The statistical relevance between the events of the two types may help to explain the breakup of the picture, and this correlation is essentially observable—for example by telephoning the station to ask if they have just been on the air. The statistical dependency itself, however, cannot be explained without reference to such theoretical entities as electromagnetic waves.

Spatio-temporal continuity obviously makes the critical difference in the two examples just mentioned. In the instance of cheating on the essay, we can provide spatio-temporally continuous processes from the common cause to the two events whose coincidence was to be explained. Having provided the continuous *causal* connections, we have furnished the explanation. In the case of trouble with the TV picture, a statistical correlation is discovered between events which are remote from one another spatially, and this correlation itself requires explanation in terms of such processes as the propagation of electromagnetic waves in space. We invoke a theoretic process which exhibits the desired continuity requirements. When we have provided spatio-temporally continuous connections between correlated events, we have fulfilled a major part of the demand for an explanation of the correlation. We shall return in a subsequent section to a more thorough discussion of the introduction of theoretic entities into explanatory contexts.

The propagation of electromagnetic radiation is generally taken to be a continuous causal process. In characterizing it as

---

[17] I realize that a full and complete explanation would require references to the theoretical aspects of perception and other psychophysiological mechanisms, but for the moment the example is being taken in common-sense terms.

continuous we mean, I suppose, that given any two spatio-temporally distinct events in such a process, we can interpolate other events between them in the process.[18] But, over and above continuity, what do we mean by characterizing a process as causal? At the very least, it would seem reasonable to insist that events that are causally related exhibit statistical dependencies. This suggests that we require, as a necessary but not sufficient condition, that explanation of statistical dependencies between events that are not contiguous be given by means of statistical relevance between neighbouring or contiguous events.[19]

We have been talking about *causes* and *causal relations*; these seem to figure essentially in the concept of explanation. The principle we are considering (as enunciated by Reichenbach) is the principle of the common *cause*; Russell's treatment of scientific knowledge relies heavily and explicitly upon *causal* relations. It seems to me a serious shortcoming of the received doctrine of scientific explanation that it does not incorporate any full-blooded requirement of causality.[20] But we must not forget the lessons Hume has taught us. The question is whether we can explicate the concept of causality in terms that do not surreptitiously introduce any 'occult' concepts of 'power' or 'necessary connection'. Statistical relevance relations represent the type of constant conjunction Hume relied upon, and spatio-temporal contiguity is also consonant with his strictures. Hume's attempt to explicate causal relations in terms of

[18] For present purposes I ignore the distinction between denseness and genuine continuity in the Cantorean sense. For a detailed discussion of this distinction and its relevance to physics, see my anthology *Zeno's Paradoxes* (Indianapolis: Bobbs-Merrill Co., 1970), especially my Introduction and the selections by Adolf Grünbaum.

[19] In the present context I am ignoring the perplexities about discontinuities and causal anomalies in quantum mechanics.

[20] In Hempel's account of deductive-nomological explanation, there is some mention of nomological relations constituting causal relations, but this passing mention of causality is too superficial to capture the features of causal processes with which we are concerned, and which seems ineradicably present in our intuitive notions about explanation.

constant conjunction was admittedly inadequate because it was an oversimplification; Russell's was also inadequate for the same reason, as I shall show in the next section. Our problem is to see whether we can provide a more satisfactory account of causal processes using only such notions as statistical relevance. We shall see in a moment that processes which satisfy the conditions of continuity and mutual statistical relevance are not necessarily causal processes. We shall, however, remain true to the Humean spirit if we can show that more complicated patterns of statistical relevance relations will suffice to do the job.

3 CAUSAL PROCESSES AND PSEUDO-PROCESSES

Reichenbach tried, in various ways, to show how the concept of causal relevance could be explicated in terms of statistical relevance. He believed, essentially, that causal relevance is a special case of statistical relevance. One of his most fruitful suggestions, in my opinion, employs the concept of a *mark*.[21] Since we are not, in this context, attempting to deal with the problem of 'time's arrow', and correlatively, with the nature and existence of irreversible processes, let us assume that we have provided an adequate physical basis for identifying irreversible processes and ascertaining their temporal direction. Thus, to use one of Reichenbach's favourite examples, we can 'mark' a beam of light by placing a red filter in its path. A beam of white light, encountering such a filter, will lose all of its frequencies except those in the red range, and the red colour of the beam will thus be a mark transmitted onward from the point at which the filter is placed in its path. Such marking procedures can obviously be used for the transmission of information along causal processes.

In the context of relativity theory, it is essential to distinguish causal processes, such as the propagation of a light

[21] Although Reichenbach often discussed the 'mark method' of dealing with causal relevance, the following discussion is based chiefly upon *The Direction of Time*, op. cit., §23.

ray, from various pseudo-processes, such as the motion of a spot of light cast upon a wall by a rotating beacon. The light ray itself can be marked by the use of a filter, or it can be modulated to transmit a message. The same is not true of the spot of light. If it is made red at one place because the light beam creating it passes through a red filter, that red mark is not passed on to the successive positions of the spot. The motion of the spot is a well-defined process of some sort, but it is not a causal process. The causal processes involved are the passages of light rays from the beacon to the wall, and these can be marked to transmit a message. But the direction of message transmission is from the beacon to the wall, not across the wall. This fact has great moment for special relativity, for the light beam can travel no faster than the universal constant c, while the spot can move across the wall at arbitrarily high velocities. Causal processes can be used to synchronize clocks; pseudo-processes cannot. The arbitrarily high velocities of pseudo-processes cannot be exploited to undermine the relativity of simultaneity.[22]

Consider a car travelling along a road on a sunny day. The car moves along in a straight line at 60 m.p.h., and its shadow moves along the verge at the same speed. If the shadow encounters another car parked on the verge, it will be distorted, but will continue on unaffected thereafter. If the car collides with another car and continues on, it will bear the marks of the collision If the car passes a building tall enough to cut off the sunlight the shadow will be destroyed, but it will exist again immediately when the car passes out of the shadow of the building. If the car is totally destroyed, say by an atomic explosion, it will not automatically pop back into existence after the blast and continue on its journey as if nothing had happened.

There are many causal processes in this physical world;

---

[22] See Hans Reichenbach, *The Philosophy of Space and Time* (New York: Dover Publications, 1957), §23, for a discussion of 'unreal sequences', which I have chosen to call 'pseudo-processes'.

among the most important are the transmission of electro-magnetic waves, the propagation of sound waves and other deformations in various material media, and the motion of physical objects. Such processes transpire at finite speeds no greater than that of light; they involve the transportation of energy from one place to another, and they can carry messages. Assuming, as we are, that a temporal direction has been established, we can say that the earlier members of such causal processes are *causally relevant* to the later ones, but not conversely.[23] Causal relevance thus becomes an asymmetric relation, one which we might also call 'causal influence'. We can test for the relation of causal relevance by making marks in the processes we suspect of being causal and seeing whether the marks are, indeed, transmitted. Radioactive 'tagging' can, for example, be used to trace physiological causal processes. The notion of causal relevance has been aptly characterized by saying, 'You wiggle something over here and see if anything wiggles over there.' This formulation suggests, of course, some form of human intervention, but that is obviously no essential part of the definition. It does not matter what agency is responsible for the marking of the process. At the same time, experimental science is built upon the idea that *we* can do the wiggling.[24] There is an obvious similarity between this approach and Mill's methods of difference and concomitant variation.

Just as it is necessary to distinguish causal processes from pseudo-processes, so also is it important to distinguish the relation of causal relevance from the relation of statistical relevance, especially in view of the fact that pseudo-processes exhibit striking instances of statistical relevance. Given the moving spot of light on a wall produced by our rotating beacon,

[23] Although Reichenbach seemed to maintain in his earlier writings, such as *The Philosophy of Space and Time*, that the mark method could be taken as an independent criterion of temporal direction (without any other basis for distinguishing irreversible processes), he abandoned that view in the later work *The Direction of Time*.

[24] We must, however, resist the strong temptation to use intervention as a criterion of temporal direction; see *The Direction of Time*, §6.

the occurrence of the spot at one point makes it highly probable that the spot will appear at a nearby point (in the well-established path) at some time very soon thereafter. This is not a certainty, of course, for the light may burn out, an opaque object may block the beam, or the beacon may stop rotating in its accustomed fashion. The same is true of causal processes. Given an occurrence at some point in the process, there is a high probability of another occurrence at a nearby point in the well-established path. Again, however, there is no certainty, for the process may be disturbed or stopped by some other agency. These considerations show that pseudo-processes may exhibit both continuity and statistical relevance among members; this establishes our earlier contention that these two properties, though perhaps necessary, are not sufficient to characterize causal processes.

Pseudo-processes exhibit the same basic characteristics as correlated events or improbable coincidences which require explanation in terms of a common cause. There is a strong correlation between the sudden drop of the barometer and the occurrence of a storm; however, fiddling with a barometer will have no effect upon the storm, and marking or modifying the storm (assuming we had power to do so) would not be transmitted to the (earlier) barometer reading. The pseudo-process is, in fact, just a fairly elaborate pattern of highly correlated events produced by a common cause (the rotating beacon). Pseudo-processes, like other cases of non-causal statistical relevance, require explanation; they do not provide it, even when they possess the sought-after property of spatio-temporal continuity.

One very basic and important principle concerning causal relevance—i.e. the transmission of marks—is, nevertheless, that it seems to be embedded in continuous processes. Marks (or information) are transmitted continuously in space and time. Spatio-temporal continuity, I shall argue, plays a vital rôle in theoretical explanation. The fact that it seems to break down in quantum mechanics—that quantum mechanics seems unavoidably to engender causal anomalies—is a source of great distress. It is far more severe, to my mind, than the discomfort we should experience on account of the apparent breakdown of determinism in that domain. The failure of determinism is one thing; the violation of causality quite another. As I understand it, determinism is the thesis that (loosely speaking) the occurrence of an event has probability zero or one in the presence of a complete set of statistically relevant conditions. Indeterminism, by contrast, obtains if there are complete sets of statistically relevant conditions (i.e. homogeneous reference classes) with respect to which the event may either happen or not—the probability of its occurrence has some intermediate value other than zero or one.[25] The breakdown of causality lies in the fact that (in the quantum domain) causal influence is not transmitted with spatio-temporal continuity. This, I take it, formulates a fundamental aspect of Bohr's principle of complementarity as well as Reichenbach's principle of anomaly.[26] Causal influence need not be deterministic to exhibit continuity; we are construing causal relevance as a species of statistical relevance. Causality, in this sense, is entirely compatible with indeterminism, but quantum mechanics goes beyond indeterminism in its admission of familiar spatio-temporal discontinuities.[27] In classical physics and relativity theory, however, we retain the principle that all causal influence is via action by contact. It is

[25] This conception of determinism, which seems to me especially suitable in the context of discussions of explanation, is elaborated in my essay 'Statistical Explanation', §4. Note also that technically it is illegitimate to identify probability one with invariable occurrence and probability zero with universal absence, but that technicality need not detain us, and I ignore it in the text of this paper.

[26] *The Direction of Time*, p. 216. See also Hans Reichenbach, *Philosophic Foundations of Quantum Mechanics* (Berkeley and Los Angeles: University of California Press, 1946).

[27] It would be completely compatible with indeterminism and causality to suppose that a 'two-slit experiment' performed with macroscopic bullets would yield a two-slit statistical distribution which is just the superposition of two one-slit patterns when large numbers of bullets are involved. At the same time, it might be the case that no trajectory of any individual bullet is precisely determined by the physical conditions. This imaginary situation differs sharply, of course, from the familiar two-slit experiment of quantum mechanics.

doubtful, to say the least, that action by contact can be maintained in quantum mechanics. Even in the macrocosm, however, pseudo-processes may display obvious discontinuities, as for example, when the spot of light from the rotating beacon must 'jump' from the edge of a wall to a cloud far in the background.

Another fundamental characteristic of causal influence is its asymmetric character; in this respect it differs from the relation of statistical relevance. It is an immediate consequence of our foregoing definition of statistical relevance that A is relevant to B if B is relevant to A.[28] This has the consequence that effects are statistically relevant to causes if (as must be the case) causes are statistically relevant to their effects. As we shall see below, Reichenbach defines the *screening-off relation* in terms of statistical relevance; it is a non-symmetric relation from which the relation of causal relevance inherits its asymmetry. The property of asymmetry is crucial, for the common cause which explains a coincidence always precedes it.

4. THEORETICAL EXPLANATION

In our world the principle of the common cause works rather nicely. We can explain the identical essays by tracing them back to a common cause via two continuous causal processes. These causal processes are constituted, roughly speaking, of events that are in principle observable, and which were in fact observed by the two plagiarists. Many authors, including Hume very conspicuously, have explained how we may endow our world of everyday physical objects with a high degree of spatio-temporal continuity by suitably interpolating observable objects and events between observed objects and events. Russell has discussed at length the way in which similar structures grouped around a centre could be explained in terms of the propagation of continuous causal influence from the

[28] See *Statistical Explanation and Statistical Relevance*, p. 55, esp. note 53.

common centre; indeed, this principle became one of Russell's postulates of scientific inference.[29] In many of his examples, if not all, the continuous process is in principle observable at any point in its propagation from the centre to more distant points at later times.

Although we can endow our world with lots of continuity by reference to observable (though unobserved) entities, we cannot do a very complete job of it. In order to carry through the task we must introduce some entities that are unobservable, at least for ordinary human capabilities of perception. If, for example, we notice that the kitchen windows tend to get foggy on cold days when water is boiling on the stove, we connect the boiling on the stove with the fogging of the windows by hypothesizing the existence of water molecules that are too small to be seen by the naked eye, and by asserting that they travel continuous trajectories from the pan to the window. Similar considerations lead to the postulation of microbes, viruses, and genes for the explanation of such phenomena as the spread of disease and the inheritance of biological characteristics. Note, incidentally, how fundamental a role the transmission of a mark or information plays in modern molecular biology. Electromagnetic waves are invoked to fulfil the same kind of function; in the explanation of the TV picture disturbance, the propagation of electromagnetic waves provided the continuous connection. These unobservable entities are not fictions—not simple-minded fictions at any rate—for we maintain that it is possible to detect them at intermediate positions in the causal process. Hertz detected electromagnetic waves; he could have positioned his detector (or additional detectors) at intermediate places. The high correlation between a spark in the detecting loop and a discharge at the emitter had to be explained by a causal process travelling continuously in space and time. Moreover, the water molecules from the boiling pan will condense on a chilled tumbler anywhere in the kitchen. Microbes

[29] Russell, *Human Knowledge*, Part VI, ch. VI. It is called 'the structural postulate'.

and viruses, chromosomes and genes, all can be detected with suitable microscopes; even heavy atoms can now, it seems, be observed with the electron scanning microscope. The claim that there are continuous causal processes involving unobservable objects and events is one that we are willing to test; along with this claim goes some sort of theory about how these intermediate parts of the process can be detected. The existence of causal relevance relations is also subject to test, of course, by the use of marking processes.

Many philosophers, most especially Berkeley, have presented detailed arguments against the view that there are unobserved physical objects. Berkeley did, nevertheless, tacitly admit the common cause principle, and consequently invoked God as a theoretical entity to explain statistical correlations among observed objects. Many other philosophers, among them Mach, presented detailed arguments against the view that there are unobservable objects. Such arguments lead either to phenomenalism (as espoused, for example, by C. I. Lewis) or instrumentalism (as espoused by many early logical positivists). Reichenbach strenuously opposed both of these views, and in the course of his argument he offers a strange analogy—namely, his cubical world. [30]

Reichenbach invites us to consider an observer who is confined to the interior of a cube in which a bunch of shadows appear on the various walls. Careful observation reveals a close correspondence between the shadows on the ceiling and those on one of the walls; there is a high statistical correlation between the shadow events on the ceiling and those on the wall. For example, when the observer notices what appears to be the shadow of one bird pecking at another on the ceiling, he finds the same sort of shadow-pattern on the wall. Reichenbach argues that these correlations should be explained as shadows of the same birds cast on the ceiling and the wall; that is, birds

[30] Hans Reichenbach, *Experience and Prediction* (Chicago: University of Chicago Press, 1938), esp. §14. Contrary to popular opinion, Reichenbach was never a logical positivist, and he regarded *Experience and Prediction* as his *refutation* of positivism

outside of the cube should be postulated. It is further postulated that they are illuminated by an exterior source which makes the shadows of the same birds appear on the translucent material of both the ceiling and the wall. He stipulates that the inhabitant of the cube cannot get to the ceiling or walls to poke holes in them, or any such thing, so that it is physically impossible for the inhabitant to observe the birds directly. Nevertheless, according to Reichenbach, he should infer their existence. [31] Reichenbach is doing precisely what he advocated explicitly in his later work; he is explaining a relation of statistical relevance in terms of relations of causal relevance, invoking a common cause to explain the observed non-contiguous coincidences. The causal processes he postulates are, of course, spatio-temporally continuous.

In *Experience and Prediction* Reichenbach claims that the theory of probability enables us to infer, with a reasonable degree of probability, the existence of entities of unobservable types. This claim seems problematic, to say the least, and I was never quite clear how he thought it could be done. One could argue that all we can observe in the cubical world are constant conjunctions between patterns on the ceiling and patterns on the wall. If constant (temporal) conjunction were the whole story as far as causality is concerned, then we could say that the patterns on the ceiling cause the patterns on the wall, or vice versa. There would be no reason to postulate anything beyond the shadows, for the constant conjunctions are given observationally, and they are all we need. The fact that they are not connected to one another by continuous causal lines would be no ground for worry; there would be no reason to postulate a common cause to link the observed coincidences via *continuous* causal processes. This, a very narrow Humean might say, is the

[31] Reichenbach does not say whether there are any birds *inside* the cube, so that the inference is to entities outside the cube quite like those on the inside, or no birds on the inside to give a clue to the nature of the inferred exterior birds. If his analogy is to be interesting we must adopt the latter interpretation and demand that the observer postulate theoretical entities quite unlike those he has observed.

entire empirical content of the situation; we cannot infer even with probability that the common cause exists. Such counter-arguments might be offered by phenomenalists or instrumentalists.

Reichenbach is evidently invoking (though not explicitly in 1938) his principle that statistical relevance must be explained by causal relevance, where causal relevance is defined by continuity and the ability to transmit a mark. In the light of this principle, we may say that there is a certain probability $P(A)$ that a particular pattern (the shadow of one bird pecking at another) will appear on the ceiling, and a certain probability $P(B)$ that a similar pattern will appear on the wall. There is another probability $P(A \& B)$ that this pattern will appear both on the ceiling and on the wall at the same time. This latter probability seems to be much larger than it would be if the events were independent, i.e.

$$P(A \& B) \gg P(A) \times P(B).$$

Reichenbach's principle asserts that this sort of statistical dependency demands causal explanation if, as in this example, A and B are not spatio-temporally contiguous. Using this principle, Reichenbach can certainly claim that the existence of the common cause can be inferred with a probability; otherwise we would have to say that the probability of $A \& B$ is *equal to* the product of the two individual probabilities, and that we were misled into thinking that an inequality holds because the observed frequency of $A \& B$ is much larger than the actual probability. In other words, the choice is between a common cause and an exceedingly improbable coincidence. This makes the common cause the less improbable hypothesis. But the high frequency of the joint occurrence is statistically miraculous only if there are no alternatives except fortuitous coincidence or a continuous connection to a common cause. If we could have causal relevance without spatio-temporal contiguity, no explanation would be required, and hence there would be no probabilistic evidence for the existence of the common cause.

If, on the other hand, we can find an adequate basis for adopting the principle that statistical relevancies must be explained by continuous causal processes, then it seems we have sufficient ground for postulating or inferring the existence of theoretical entities.

In rejecting the notion that we have an impression of necessary connection, Hume analysed the causal relation in terms of constant conjunction. As he realized explicitly, his analysis of causation leaves open the possibility of filling the spatio-temporal gaps in the causal chain by interpolating events between observed causes and observed effects. In so doing, he maintained, we simply discover a larger number of relations of constant conjunction with higher degrees of spatio-temporal contiguity. In recognition of the fact that causal relations often serve as a basis for inference, Hume attempts to provide this basis in the 'habit' or 'custom' to which observed constant conjunction naturally gives rise.

Russell has characterized causal lines as continuous series of events in which it is possible to infer the nature of some members of the series from the characteristics of other events in the same series. This means, in our terms, that there are relations of statistical relevance among the members of such series. Although causal series have enormous epistemological significance for Russell, providing a basis for our knowledge of the physical world, his characterization of causal series is by no means subjective. It is by virtue of factual relations among the members of causal series that we are enabled to make the inferences by which causal processes are characterized.

Statistical relevance relations do provide a basis for making certain kinds of inferences, but they do not have all of the characteristics of causal relevance as defined by Reichenbach; in particular, they do not always have the ability to transmit a mark. Although Russell did not make explicit use of mark transmission in his definitions, his approach would seem hospitable to the addition of this property as a further criterion of causal processes. Russell emphasizes repeatedly the idea that perception is a causal process by which structure can be transmitted. He frequently cites processes like radio transmission as physical analogues of perception, and he obviously considers such examples extremely important. The transmission of messages by the modulations of radio waves is a paradigm of a mark. In similar ways, the absorption of all frequencies but those in the green range from white light falling upon a leaf is a suggestive case of the marking of a causal process involved in human perception. The transmitted mark conveys information about the interaction that is responsible for the mark. The mark principle thus seems to me to be a desirable addition to Russell's definition of causal processes, and one that can be fruitfully incorporated into his postulates of scientific knowledge.

I do not wish to create the impression that ability to transmit a mark is any mysterious kind of necessary connection or 'power' of the sort Hume criticized in Locke. Ability to transmit a mark is simply a species of constant conjunction. We observe that certain kinds of events tend to be followed by others in certain kinds of processes. Rays of white light are series of undulatory events which are spatio-temporally distributed in well-defined patterns. Events which we would describe as passage of light through a red filter are followed by undulations with frequencies confined to the red range; undulations characterized by other frequencies do not normally follow thereupon. It is a fact about this world (at least as long as we stay out of the quantum realm) that there are many continuous causal processes that do transmit marks. This is fortunate for us, for such processes are highly informative. Russell was probably right in saying that without them we would not have anything like the kind of knowledge of the physical world we actually do have. It is not too surprising that causal processes capable of carrying information figure significantly in our notion of scientific explanation. To maintain that such processes are continuous, we must invoke theoretical entities.

Let us then turn to the motivation for the continuity requirement.

5 SPATIO-TEMPORAL CONTINUITY

Throughout this paper we have been discussing the continuity requirement on causal processes; it is now time to see why they figure so importantly in the discussion. If special relativity is correct, there are no essential spatio-temporal limitations upon relations of statistical relevance, but there are serious limitations upon relations of causal relevance. Any event A that we choose can be placed at the apex of a Minkowski light cone, and this cone establishes *a cone of causal relevance*. The backward section of the cone, containing all of the events in the absolute past of A, contains all events that bear the relation of causal relevance to A. The forward part of the light cone, which contains all events in the absolute future of A, contains all events to which A may bear the relation of causal relevance. In contrast, an event B which is in neither the backward nor the forward sections of the cone cannot bear the relation of causal relevance to A, nor can A bear that relation to B. Nevertheless, B can sustain a relation of *statistical* relevance to A. When this occurs, according to Reichenbach's principle, there must be a common cause C somewhere in the region of overlap of the backward sections of the light cones of A and B. The relation of statistical relevance is *not* explainable, as mentioned above, by a common effect in the region of overlap of the forward sections of the two light cones.[32]

If our claims are correct, any *statistical* relevance relation between two events can be explained in terms of *causal* relevance relations. Causal relevance relations are embedded in continuous causal processes. If, therefore, an event C is causally relevant to A, then we can, so to speak, mark off a boundary in

[32] These statements obviously represent factual claims about this world. We believe they are true, and if they are true they are very important. But we have no reason to think they are true in all possible worlds.

the backward part of the light cone (i.e. the causal relevance cone) and be sure that C is either within that part of the cone or else that it is connected with A by a continuous causal process which crosses that boundary. Hence, to investigate the question of what events are causally relevant to A, we have only to examine the interior and boundary of some spatial neighbourhood of A for a certain time in the immediate past of A. We can thus ascertain whether such an event lies within that neighbourhood, or whether a connecting causal process crosses the boundary. We have been assuming, let us recall, that a continuous causal process can be detected anywhere along its path. This means that we do not have to search the whole universe to find out what events bear relations of *causal* relevance to A.[33]

If we make it our task to find out what events are *statistically* relevant to A, all of the events in the universe are potential candidates. There are, in principle, no spatio-temporal limitations on statistical relevance. But, it might be objected, statistical relevance relations can serve as a basis for inductive inference, or at least for inductive behaviour (for example, betting). How are we therefore justified, if knowledge is our aim, in restricting our considerations to events which are causally relevant? The answer lies in the *screening-off* relation.[34]

If A and B are two events that are statistically relevant to one another, but neither is causally relevant to the other, then there must be a common cause C in the region of overlap of the past light cones of A and B. It is possible to demonstrate the causal relevance of C to A by showing that a mark can be transmitted along the causal process from C to A, and the causal relevance of C to B can be demonstrated in a similar fashion. There is, however, no way of transmitting a mark from B to A or from A to B. When we have that kind of situation,

[33] In this connection it is suggestive to remember Popper's distinction between falsifiable and unfalsifiable existential statements.

[34] Reichenbach, *The Direction of Time*, p. 189, and Salmon, 'Statistical Explanation', §7.

which can be unambiguously defined by the use of marking techniques, we find that the statistical relevance of B to A is absorbed in the *statistical* relevance of C to A. That is just what the screening-off relation amounts to. Given that B is statistically relevant to A, and C is statistically relevant to A, we have

$$P(B, A) > P(A) \text{ and } P(C, A) > P(A).$$

To say that C screens off B from A means that, given C, B becomes statistically irrelevant to A; i.e.

$$P(B \,\&\, C, A) = P(C, A).$$

Thus, for example, though the barometer drop indicates a storm and is statistically relevant to the occurrence of the storm, the barometer becomes statistically irrelevant to the occurrence of the storm given the meteorological conditions which led to the storm and which are indicated by the barometer reading. The claim that statistical relevance relations can always be explained in terms of causal relevance relations therefore means that causal relevance relations screen off other kinds of statistical relevance relations.

The screening-off relation can be used, moreover, to deal with questions of causal proximity. We can say in general that more remote causal relevance relations are screened off by more immediate causal relevance relations. Part of what we mean by saying that causation operates via action by contact is that the more proximate causes absorb the entire influence of more remote causes. Thus, we do not even have to search the entire backward section of the light cone to find *all* factors relevant to the occurrence of A. A *complete* set of factors statistically relevant to the occurrence of a given event can be found by examining the interior and boundary of an appropriate neighbouring section of its past light cone. Any factor outside of that portion of the cone that is, by itself, statistically relevant to the occurrence of the event in question is screened off by events within that neighbouring portion of the light cone. These are strong factual claims; if correct, they have an enormous bearing upon our conception of explanation.

6 CONCLUSIONS

In this paper I have been trying to elaborate the view of scientific explanation which is present, at least implicitly I think, in the works of Russell and Reichenbach. Such explanation is causal in a very deep and pervasive sense; yet I believe it does not contain causal notions that have been proscribed by Hume's penetrating critique. This causal treatment accounts in a natural way for the invocation of theoretical entities in scientific explanation. It is therefore, I hope, an approach to scientific explanation that fits especially well with scientific realism (as opposed to instrumentalism). Still, I do not wish to claim that this account of explanation establishes the realistic thesis regarding theoretical entities. An instrumentalist might well ask: Is the world understandable because it contains continuous causal processes, or do we make it understandable by imputing continuous causal processes? This is a difficult and far-reaching query.

It is tempting to try to argue for the realist alternative by saying that it would be a statistical miracle of overwhelming proportions if there were statistical dependencies between remote events which reflect precisely the kinds of dependencies we should expect if there were continuous causal connections between them. At the same time, the instrumentalist might retort: what makes remote statistical dependencies any more miraculous than contiguous ones? Unless one is willing to countenance (as I am not) some sort of pre-Humean concept of power or necessary connection, I do not know quite what answer to give. We may have reached a point at which a pragmatic vindication, a posit, or a postulate is called for. It may be possible to argue that scientific understanding can be achieved most efficiently (if such understanding is possible at all), by searching for spatio-temporally continuous processes capable of transmitting marks. This may be the situation with which Russell attempted to cope by offering his postulates of scientific inference.[35] The preceding section was an attempt to spell out the methodological advantages we gain if the world is amenable to explanations of this sort, but I do not intend to suggest that the world is otherwise totally unintelligible. After all, we still have to cope with quantum mechanics, and that does not make scientific understanding seem hopeless.

Regardless of the merits of the foregoing account of explanation, and regardless of the stand one decides to take on the realism-instrumentalism issue, it is worthwhile, I think, to contrast this account with the standard deductive-nomological account. According to the received view, empirical laws, whether universal or statistical, are explained by deducing them from more general laws or theories. Deductive subsumption is the key to theoretical explanation. According to the present account, statistical dependencies are explained by, so to speak, filling in the causal connections in terms of spatio-temporally continuous causal processes. I do not mean to deny, of course, that there are physical laws or theories which characterize the causal processes to which we are referring—laws of mechanics which govern the motions of material bodies, laws of optics and electromagnetism that govern the propagation of electromagnetic waves, etc. The point is, rather, that explanations are not arguments on this view. Causal or theoretical explanation of a statistical correlation between distinct types of events is an exhibition of the way in which those regularities fit into the causal structure of the world—an exhibition of the causal connections between them which give rise to the statistical relevance relations.

---

[35] I have discussed Russell's views on his postulates in some detail in 'Russell on Scientific Inference *or* Will the Real Deductivist Please Stand Up', in George Nakhnikian, ed., *Bertrand Russell's Philosophy* (London: Gerald Duckworth & Co., 1974), pp. 183–208. In the same essay I have discussed aspects of Popper's methodological approach that are relevant to this context.

Comment

BY D. H. MELLOR

Professor Salmon agrees 'that *causal relevance* (or causal influence) plays an indispensable rôle in scientific explanation, and . . . attempt[s] to show how this relation can be explicated in terms of the concept of statistical relevance' (pp. 120-1). I wish to indicate three main obstacles I find to my accepting his explication. The first is a general objection to any explication of causation in terms of statistical relevance as understood by Salmon. The second more specifically concerns the role of Reichenbach's 'mark' in Salmon's explication. The third has to do with his explication of the asymmetry of the cause-effect relation.

1. For Salmon, statistical relevance and statistical explanation are notions applying essentially to kinds or classes of events rather than to particular events. Causal relevance and causal explanation, however, seem to me essentially tied to particular events. No doubt, if an event e causes an event e', it does so by virtue of certain features of the two events. To take one of Salmon's examples, it is the feature of being a transmission that *inter alia* makes a police transmission cause a herringbone pattern on my TV screen. Now it *might* follow that *any* pair of events with all the relevant features will be similarly related as cause to effect; and for a Humean that general fact might suffice to explain away the otherwise mysterious cause-effect relation. It is of course disputed whether any such 100 per cent generalization follows from a singular causal statement, let alone whether it will serve to explicate the statement. But the plausibility of supposing some such generalization to be involved makes it natural to speak elliptically of the causal relation holding between classes or kinds of events, namely those having the features that make e cause e'. What is meant is that each event in the one class causes the corresponding event in the other.

Salmon wants us to take these as merely extreme cases of a more general statistical relation of causal influence, where the correlation between events of the two kinds may be less than 100 per cent. I find, however, that I can make no sense of explicating the cause-effect relation in terms of a less than 100 per cent correlation. All Salmon's examples seem to me to appeal only when construed as 100 per cent cases. It may indeed be, to pursue his example, that 'when I find a strong statistical correlation [say 95 per cent] between the operation of the police transmitter and the breakup of the picture, I conclude that the police broadcast is part of the explanation of the television malfunction' (p. 123). I may even reasonably conclude with Salmon that 'the operation of the police transmitter is the cause (or a part of the cause) of the bad TV picture' (ibid.). But I should then expect the occasions when the transmission was followed by *no* pattern to be explainable either by the absence of some other condition (like my set not being switched on) or of some relevant feature of the transmission itself (like not being on some particular wavelength). Either way, if I am prepared to generalize my singular causal complaint at all, as Salmon does, it will be to some 100 per cent causal correlation. The original 95 per cent correlation, which made me (rapidly and no doubt reasonably) suspicious of the police transmission, reappears as a merely accidental mixture of (for example) transmissions on this and that wavelength. The proportion of transmissions on the offending wavelength could have been anything, down to 5 per cent or less, without making those transmissions any less the cause of my TV pattern. The lower this accidental proportion, the slower I shall be to guess the pattern's cause, that's all.

If there is supposed to be *only* the 95 per cent correlation, however, I confess I do not see how it serves to explicate the supposed causal connection. For now we are invited to speak of police transmissions *causing* the pattern in an undifferentiated class of cases in which sometimes there is a transmission and no pattern and sometimes a pattern and no transmission. Yet if the literature of causation agrees on anything, it is surely that *e* can only cause *e'* if both *e* and *e'* at least *occur*.

One might say that a 95 per cent correlation shows or suggests at any rate that those transmissions which *are* followed by patterns cause them, and it may of course be true that they do. But explicating the causal connection in these cases at least involves differentiating them from the others, however few, in which the pattern does *not* follow (or the transmission does not precede), so that no causal connection can obtain between them. It is this differentiation that any correlation less than 100 per cent fails to provide. 'You wiggle something over here and see if anything wiggles over there' may be, as Salmon says (p. 131), an apt characterization of causal relevance.

I am not claiming that a 100 per cent correlation is necessary for a singular cause-effect relation to obtain, still less that it is sufficient. For all I care here, a transmission may cause a TV pattern on a particular occasion even if there is no better than a 5 per cent correlation between respectable classes of such events. If so, Hume's account of causation is wrong. I am not sure enough of the proper analysis of the non-material conditionals involved to rule out that possibility. But I am sure enough to rule out the merely statistical analysis that makes the truth of (for example)

'If the transmission had not occurred, other things being equal, the pattern would not have occurred'

compatible with the pattern occurring without the transmission on a particular occasion on which other things *were* equal.

2. My second difficulty concerns Salmon's use of 'marking' (p. 129f) as a test for causal processes. Marking seems to me merely a special case of causation, not a more primitive notion in terms of which causation can usefully be explicated.

Consider how marking might be used to test the causality of a process. (To avoid raising again the problems of §1 above, I consider only deterministic cases, but what follows would apply equally to the other cases if they made causal sense.) The question is whether an earlier event e in the process causes a later one e'. The particular process we observe leaves us in doubt, so we examine another of the same kind. This other is the same in the features that are suspected of causal significance, but we make the earlier of the corresponding pair of events differ in some suitable respect from e and see if the later differs similarly from e'. In Salmon's examples we wonder (p. 129) if the light passing A is in the causal ancestry of (i.e. upstream in the same beam as) the light passing B a suitably small time later. Or (p. 131) we wonder (for example) if this input to an organism is in the causal ancestry of that output. So in the first case we send some more light, of a different colour, past A and see if its supposed descendant at B acquires the new colour; in the second we make otherwise similar input radio-active and see what happens to the ensuing output.

In each case we observe that an event e_1 suitably like e, but differing in having some marking features F, is followed by an event e_1' suitably like e', which also differs in having the feature F. But so what? It might just be a coincidence that e_1 being F is followed by e_1' being F; in which case we have no reason to infer that e_1 and e_1' (nor hence e and e') are part of a causal, rather than merely a 'pseudo-process' (Salmon, p. 130). Evidently we need to show that e_1 being F causes e_1' to be F, i.e. that the process linking them is causal. But that is the very question which marking e with the feature F was supposed to answer.

It will obviously not help to pile marks upon marks, to polarize the light at A as well as making it red or to make the organism's food luminous as well as radioactive. The same question merely repeats itself. In each case we can answer the question by applying such customary simple-minded tests as doing it all again under the same conditions and seeing if we always get the same result. If we do, we may convince ourselves that the constant conjunction (Fe_i & Fe_i', $i = 1, 2, \ldots$) is no coincidence but genuinely marks a cause-effect relation between

e_i and e_i'. But then we could have applied the same tests to the original white light, the original undoctored food, without appealing to marks at all. Marking is in principle redundant; if we cannot detect causal processes without marking, then in particular we cannot detect marking.

No one would deny that marking has practical virtues, in giving events more readily observed features, so that their causal connections may be more readily tested for. But I can see no more to it than that. The idea that there is more to marking I suspect comes from attaching specious significance to the distinction between an event occurring at all and it having some specified feature. Being a cause is seen as a matter of occurring at all, while marking is seen as a matter of having a suitable feature. Thus light passing A causes light to pass B because, *inter alia*, if there were no light at A no light would follow at B; the colour of the light at B marks its colour at A because, *inter alia*, if the light at A is changed in colour so is that at B.

If this is supposed to be the distinction between cause and mark it will not do. Suppose we say that e being the cause of e' means in part that had e not occurred e' would not have occurred. What we really mean is that had not the relevant features of e not been instantiated then and there, nor would those of e' have been instantiated. Normally the relevant features of e and e' are not explicitly specified: they are understood to be among those given in the descriptions used to refer to e and e'. This I think is because we are less apt to identify events under different descriptions than we are so to identify things. We therefore less readily think of e and e' as having features other than those by which we refer to them, which they could be conceived to retain even if deprived of whatever makes e cause e'. So our causal conditionals typically hypothesize only the non-occurrence of the events concerned, rather than their existing but lacking their causally significant features.

I reckon it is for the same reason that we naturally speak of explaining an event *tout court*, where we rather speak of explaining a thing having this or that feature. The present point is that such talk of explaining events is always elliptical: we mean by it not the explaining merely of an occurrence, totally unspecified, but of an event of this or that kind, the kind being tacitly given in the course of referring to the event. Thus, as we have noted already, it is the feature of being a transmission that makes a police transmission cause a herringbone pattern on my TV screen. And what reference to the transmission causally explains is the feature the pattern has of differing from those put out officially at that time on the channel my set is tuned to.

Typically, then, in causal explanation, the relevant features of the events concerned are specified tacitly. In talk of marking, on the other hand, the relevant features are specified explicitly. The events in the process, with their causal relations, are conceived as occurring independently of the marking by which they acquire this or that extra feature. It is thus temptingly easy to think of marking as a process both different in kind from, and a universally applicable test for, causal influence, by reference to which causal influence can be explicated.

It seems to me important to resist this temptation for practical as well as theoretical reasons. Once we see that cause-effect relations result from features of events rather than from their bare occurrence, we can see that e being F_1 may cause e' to be F_1' without e being F_2 causing e' to be F_2'. In particular, marking (where typically $F_i = F_i'$) is *not* an infallible test of a process being causal in any other respect. It is, of course, a trivially infallible test of a process being causal in respect of the marking feature F, but no more. Light being red at A may cause light to be red at B without light at A causing light at B. (Suppose, for example, colour at A causes the independent source of light at B to be filtered to the same colour.) Radioactivity in an organism's input may cause radioactivity in its output without the input being the same stuff as the output or being causally responsible for any other feature of it.

In conclusion, therefore, I find that marking, as a merely special case of causation, cannot play the role in its explication that Salmon assigns to it.

3. Salmon explicates the essential asymmetry of the cause-effect relation in terms of that of the temporal 'later than' relation: 'Assuming . . . that a temporal direction has been established, we can say that the earlier members of such causal processes are *causally relevant* to the later ones, but not conversely. Causal relevance thus becomes an asymmetric relation, one which we might also call "causal influence"' (p. 131). I do not of course, dispute the intimate and necessary connection between the asymmetries of the cause-effect and 'later than' relations; what I doubt is that the asymmetry of the 'later than' relation is the more fundamental. If asked to explain why 'later than' is an asymmetrical relation, and how it differs from 'earlier than', I fear I should have to appeal *inter alia* to the asymmetry of cause and effect. If so, Salmon's account of the latter asymmetry in terms of the former involves a circular tour of uncomfortably small radius. (The circle would be broken if there were, as Salmon has suggested to me Reichenbach provides in *The Direction of Time*, ch. 4, an atemporal criterion to distinguish a common cause of two events from a common effect of them; but I cannot see that there is.)

Comment

BY L. JONATHAN COHEN

In his very interesting and stimulating paper Professor Salmon advances two main criticisms of what he calls the received theory of scientific explanation—which for present purposes we can treat as Hempel's theory. The first criticism is that it ignores the need for the explanans to be causally relevant to the explanandum. By ignoring this it makes a man's failure to take his wife's birth control pills explain his failure to become pregnant. The second criticism is that *a fortiori*, as it were, the received theory of explanation ignores the need for any causal connection between explanans and explanandum to be a continuous process. The boiling kettle explains the fogginess of the windows only if water molecules can be supposed to travel from the kettle to the window. But Salmon is doing something more in his paper than just making these two criticisms of the Hempelian theory of explanation. He is also claiming that each of the two requirements for a good explanation which are neglected by the Hempelian theory is needed because certain statistical relevance relations obtain in the situations concerned. All deductive-nomological explanations, as Hempel called them, should be incorporated along with statistical explanations under what Salmon has called 'the statistical-relevance model'. First, the explanans does not have to give a high probability to the explanandum but does have to be statistically relevant to it. Secondly, what Reichenbach called 'the screening-off relation' has to be used to deal with questions of causal proximity (p. 143).

I shall deal separately with the two issues that Salmon raises. I shall argue first that a rather stronger case can be made out for the statistical-relevance model than Salmon himself allows in the recent book (*Statistical Explanation and Statistical Relevance*) to which he makes frequent reference in his paper. And I shall argue secondly that there is a good deal weaker case than Salmon claims for supposing that the screening-off relation has something important to do with the spatio-temporal continuity of causal processes.

First, then about the statistical relevance model. Salmon's paper today gives the impression that he thinks of it as requiring the posterior probability of the explanandum (the probability of the explanandum on the explanans) to be greater than its prior probability. But according to his book, this is not so. For instance, he says (p. 64), consider a mixture

of uranium 238 atoms, which have a rather long half-life
($4 \cdot 5 \times 10^9$ years), and polonium 214 atoms, which have
a rather short one ($1 \cdot 6 \times 10^{-4}$ seconds). The probability of
disintegration of an unspecified atom in the mixture is
somewhere between that for atoms of uranium 238 and
polonium 214. Now suppose that within some small specified
time interval a decay occurs. There is a high probability
of a polonium atom's disintegrating within that interval,
but a very low probability for a uranium atom. Nevertheless,
says Salmon, a given disintegration might be of a uranium
atom, so an explanation of why this atom in the mixture
disintegrated so soon would involve a transition from a prior
reference-class (the mixture) within which there was a relatively
high probability of such a disintegration to a posterior reference-
class (the uranium atoms) within which there was a relatively
low one.

It might therefore be tempting, wrote Salmon (p. 65),
to suppose that an explanation must result merely in some
change—an increase *or* a decrease—in the transition from the
prior to the posterior probability. But his model of explanation
does not even require a difference of probability here. Suppose
we play a game of coin-tossing, he says, with three coins, of
which two are biased and one is not. One of the coins has a
$0 \cdot 9$ probability of falling heads, one a $0 \cdot 1$ probability, and the
third $0 \cdot 5$ probability. The coins are brought in and out of play
in some random fashion (for example by the toss of another
unbiased coin). So the overall, prior probability of heads in the
game is $0 \cdot 5$ and if a particular coin that falls heads turns out to
be the unbiased one its posterior probability of falling heads, on
the basis of this information, is also $0 \cdot 5$. In such a case the
transition from the prior to the posterior probability involves
neither increase nor decrease. The explanation has occurred,
according to Salmon, simply in virtue of the partition of the
original, prior non-homogeneous reference-class and the
consequential assignment of the explanandum to a statistically
homogenous reference-class.

But these examples are not very persuasive, if the right distinctions are to be drawn between importantly different kinds of explanation. These examples seem to me to be examples of explanations not of why a particular event *did* happen, but of how it *could* happen. First we identify the nature of the atomic disintegration by assigning it to the broadest homogeneous reference-class and then we explain its possibility by pointing to its non-zero probability within this reference-class. To explain why it *did* occur we should need to be able to cite a favourably relevant probability of occurrence for it within the broadest homogeneous reference-class. If the transition from the prior, mixed, non-homogeneous reference-class lowers the probability but does not altogether eliminate it, we have explained how the atom could still disintegrate so soon even though it was uranium 238. But we might need to know a lot more about atoms than we do now to explain why the disintegration did actually happen. Similarly with the coin. We have explained how it *could* fall heads, but to explain why it *did* fall heads we should need to assign the coin-toss to the broadest homogeneous reference-class in which it had a favourably relevant probability of occurrences—for example to the class of tosses by skilled heads-tossers.

Indeed, if, as Salmon claims in his paper (p. 120), causal relevance plays an indispensable rôle in scientific explanation, it is difficult to see how he can explicate this fact without requiring that an increase of probability ensue as a result of transition from a prior non-homogeneous reference-class to a posterior homogeneous one. Salmon rightly insists that, *pace* Hempel, the explanandum need not have a *high* probability on the explanans. But, *pace* Salmon, this probability must be higher than the prior one. For in effect what we are doing here, when we require a *favourably* relevant probability, is to generalize the method of difference, which is intimately connected with the establishment of causality. When we apply the (probabilistically generalized) method of difference and learn that $p[S, Q \& R] > p[S, Q \& \text{not} - R]$, we implicitly learn also thereby that

$p[S, Q \& R]>p[S, Q]$. This is just a matter of mathematics. That is to say, a favourably relevant posterior probability—probability of the explanandum on the explanans—is a necessary condition for the explanans to be causally relevant to the explanandum—if by 'causally relevant' here we are not to mean something rather odd and idiosyncratic and unenlightening. In his paper (p. 131) Salmon himself mentions the similarity between his approach and Mill's methods of difference and concomitant variations. But this is not consistent with what he said in his book.

I think therefore that we have to discount what Salmon says in his book about the uranium and polonium atoms, and the game with the biased and unbiased coins, as a momentary aberration from the true doctrine of the statistical relevance model. It is then clear that this doctrine imputes a thoroughly Baconian structure to the characteristic type of argument that supports a statistical explanation. The method of difference, in a generalized form, is needed to establish relevance within the broadest homogeneous reference-class, and the method of agreement is in any case needed to establish the homogeneity of this reference-class.

I want to turn now to the second main issue that Salmon's paper raises. So far I have been arguing that a rather stronger and cleaner-cut case can be made out for the statistical relevance model than Salmon himself allows. I intend now to argue that there is a good deal weaker case than Salmon claims for supposing that the screening-off relation has something important to do with the philosophical analysis of scientific explanation.

What Salmon, following Reichenbach, means by the screening-off relation is this. Screening-off occurs when the statistical relevance of Q to S is absorbed in the statistical relevance of P to S. Given $p[S, Q]>p[S]$ and $p[S, R]>p[S]$, to say that R screens off Q from A is to say that, given R, Q becomes statistically irrelevant to S, i.e. $p[S, Q \& R]=p[S, R]$. To quote Salmon's example, though the barometer drop indicates a storm and is statistically relevant to the occurrence of the

storm, the barometer becomes statistically irrelevant to the occurrence of the storm given the meteorological conditions which led
to the storm and which are indicated by the barometer reading.

Now why should this have anything to do with the spatio-
temporal continuity of causal process? Salmon's argument
seems to run something like this. For any given event S that
has to be explained we can imagine a cone of causal relevance
with its apex at S. The section of the cone below this apex
contains all events in the absolute past of S and thus all events
causally relevant to S. The screening off relation then enables
us to distinguish those cases (like the barometer and the storm)
where statistical relevance is not a sign of causal relevance. If
S and Q are statistically but not causally relevant to one another
then there must be another event R which is a common cause of
both S and Q. So we can imagine two overlapping cones like
this

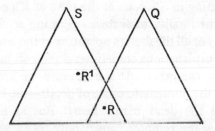

with R occurring somewhere inside the area of overlap. In
these circumstances, says Salmon, R will screen off Q from
S. But also, he says, more remote causal relevance relations are
screened off by more immediate ones, for example, presumably
R by R'. 'Part of what we mean', he claims (p. 143), 'by saying
that causation operates via action by contact is that the more
proximate causes absorb the entire influence of more remote
ones. So we don't have to search the entire lower section of
the cone to find all factors relevant to the occurrence of S.'
'A *complete* set of factors statistically relevant to the occurrence
of a given event can be found by examining the interior and
boundary of an appropriate neighbouring section' of the cone.

Salmon seems to be claiming that spatio-temporal continuity of causation from any point in S's causal past is guaranteed by the possibility of intercalating a screening off factor between that point and S. But the mere definition of the concept of screening off certainly does not ensure the existence in the real world of such intermediate events. If there are in fact any forms of action at a distance, we shall surely be disappointed in some cases when we look for proximate causes to screen off distant ones. So the most that could be hoped for from the concept of screening-off is that some appropriate methodological requirement could be formulated in terms of it. Instead of saying 'Always look for intermediate causes', perhaps it will suffice to say 'Always look for screening off factors'. But that won't do, either, since screening off can operate in either direction, as it were. The totality of factors operating in S's cone at the time of R'''s occurrence can certainly screen off any or all the factors operating in S's cone at the time of R''s occurrence. But so too can the totality of factors operating at the earlier time screen off any or all the factors operating at the later time. And in practice our explanations sometimes screen off in one direction, sometimes in the other. At the autopsy we are interested in determining the immediate cause of death—the bullet, perhaps, that entered the dead man's heart. But at the inquest we want to know the remoter cause: how was a gun fired at his heart?

What then is left of Salmon's theory here? He claims to have discerned a defect in the received, Hempelian account of scientific explanation from which his own statistical-relevance model, with its concept of screening off, does not suffer. But so far as I can see it just is not possible to specify the constraint which Salmon has in mind in a way that relies on nothing but the screening off relation and does not specify the direction in which it is to operate. We can say, if we wish, that R may explain S only if this does not involve action at a distance, and then elucidate this further by saying that if R is not proximate to S (whatever this may mean) there must be an intermediate factor or set of factors R' which has the same degree of causal relevance to S as R has. The concept of screening off *might* then be used to precisify this notion of 'same degree of causal relevance as'. But in any case the crux of the matter lies in the problem where and when and why we should suppose action at a distance impossible and *that* problem is not solved in any way by the statistical-relevance model. Perhaps it is a defect in the received, Hempelian theory that it does not discount causation at a distance. Or perhaps rather we need to distinguish between those constraints on scientific explanation which are very deeply rooted and are usefully regarded as belonging to the logical structure of explanation and those constraints, like the exclusion of action at a distance, which are somewhat less deeply rooted and may better be conceived as depending on currently accepted physical theory.

It follows that Salmon's two criticisms of the received, Hempelian theory of explanation are not on a par with one another. The first criticism—once the inconsistency in Salmon's account has been eliminated—is a valid objection to Hempel's claim that in statistical explanation the probability of the explanandum on the explanans must be a high one. If Salmon's Baconian analysis is right, as I think it is, then Hempel's account contains a definite mistake that bars many good explanations and admits many bad ones. But the second criticism urges merely a supplementation of Hempel's analysis, to rule out action at a distance, and it offers a partial elucidation of this supplement in terms of screening off. If Salmon is right here, Hempel's account is inadequate. It admits explanations it should not admit. But in this respect at least it does not bar any explanations it should not bar. And it is open to Hempel to defend himself quite plausibly on the ground that the merit of the proposed additional requirement depends rather on currently accepted physical theory than on the analysis of logical structure.

Reply to Comments

BY WESLEY C. SALMON

I   REPLY TO MR. COHEN

I am grateful to Mr. Cohen for his serious attempt to figure out what I was driving at, and for his (largely successful) efforts to clarify the fundamental issues. He has focused upon two points.

1. Although he expresses general agreement with some features of the statistical-relevance model of explanation, he finds it impossible to accept the idea that an event can be explained by showing that it has a probability less than (or even in some cases, equal to) its probability on antecedent evidence. He attributes this admittedly counter-intuitive view to a temporary mental aberration. Although one can only thank one's friends for attempting to protect them from the consequences of such episodes, I fear I must persist in this apparent folly. Instead of accepting Cohen's very appealing gloss (that cases in which the probability is unchanged or lowered in light of the explanans be viewed as explanations, not of the occurrence of the event, but of its mere possibility), I still wish to maintain, as I was arguing implicitly in the third paragraph of 'Theoretical Explanation', that an explanation of a particular event may do just that—render it less probable, or no more probable, than it was in the absence of the explanatory facts.

In that context, it will be recalled, I argued that the shift in emphasis from the evaluation of a particular explanation to the explanatory value of an entire 'theory' is important 'because it removes any temptation to suppose that we cannot explain Johnny's behaviour unless we can cite conditions in relation to which it is highly probable'. The very same argument *should* also serve to remove *the temptation to suppose that we cannot explain Johnny's behaviour unless we can cite conditions in relation to which it is* MORE *probable than before*, even though that probability may not be very high. For, it seems to me, when we ask why *this* boy (Johnny Jones) is delinquent, an adequate explanation of that fact will, by implication, be an answer to the question of why any arbitrarily selected member of the class of delinquent boys exhibits that attribute. Theoretical concerns are, I believe, fundamentally general, though practical concerns are often quite particular. Thus, to explain Johnny's delinquency, I maintain, one must provide a partition of the entire class of boys in terms of (ideally all) factors statistically relevant to the occurrence of delinquency, along with the associated probability value for each subset of the partition. In addition, of course, Johnny must be referred to the compartment to which he belongs. But the fact that he is referred to a compartment in which the probability of delinquency is low—indeed, in which it is lower than the overall probability in the original non-homogeneous class of boys, or in which it has the same value as it had in the original class of boys—does not detract from the value of that explanation. If an explanation of delinquency makes delinquency more probable for some classes of boys, it *must* also render it less probable for other classes of boys.

What we would like, of course, is a set of conditions both necessary and sufficient for the occurrence of delinquency in the original class of boys. Such conditions would provide a partition of that class into two subclasses: a subclass C_1 in which *no* member becomes delinquent, and a subclass C_2 in which *every* member becomes delinquent. We are, at present, very far from knowing of any such partition, and we cannot really guarantee that such a partition exists even in principle.[1] In the absence of this limiting sort of homogeneous partition, the next best thing would seem to be a partition of the reference class into two compartments, in one of which the attribute delinquency *almost* never occurs, and in the other of which it

[1] Even though we would be well on the road to understanding delinquency in general, and that of Johnny Jones in particular, if we knew of such necessary and sufficient conditions, I do not mean to suggest that we would, *ipso facto*, have a complete explanation, for as I argue in my paper, there may still remain important questions about the causal relations underlying the statistically homogeneous partition of the reference class.

almost always occurs. Were such a partition known, it would serve as a statistical analogue of necessary and sufficient conditions.[2] And this analogy can, I believe, be extended plausibly and fruitfully to any partition of a reference class into homogeneous subclasses.

There is considerable danger that disputes of the present sort may degenerate into fights about who gets possession of the honorific term, 'explanation', or 'explanation *simpliciter*' as opposed to 'explanation of how-possibly'. I have no desire to participate in any sort of terminological imperialism, and in company with Hempel, I would be glad to adopt a more neutral expression such as 'systematization'. But there are two points of substance I am anxious to emphasize.

First, whether it be called 'explanation', 'systematization', or by some other name, there is an almost universally acknowledged desideratum of science, closely connected with such cognitive aims as knowledge and understanding, which has often been referred to as *scientific explanation*. Under the extraordinarily persuasive influence of such writers as Braithwaite, Hempel, Nagel, and Popper, we have been led to believe that this thing, whatever you may call it, takes the form of an argument. It seems to me that the influence has been so pervasive that the identification of explanations with arguments (fulfilling certain conditions) has almost the status of a primitive intuition for many of us—an intuition whose rejection is difficult to entertain, much less to achieve. But I think this fundamental idea should be examined carefully—not accepted as a pre-analytic condition of adequacy. Moreover, I believe it should be rejected. The case for rejection has been admirably elaborated by Richard Jeffrey.[3]

I am strongly inclined to believe that it is this intuitive dentification of explanations with arguments that leads one

[2] This point is explained in Salmon, et al., *Statistical Explanation and Statistical Relevance* (Pittsburgh: University of Pittsburgh Press, 1971), p. 61.

[3] 'Statistical Explanation vs. Statistical Inference', reprinted in *Statistical Explanation and Statistical Relevance*.

to say, in the context of inductive or statistical explanation, 'Well, if the explanation cannot succeed in rendering the explanandum event highly probable, the least we can ask is that it be shown to be more probable by virtue of the explanans.' I do not mean to be dogmatic in the assertion, in the opening paragraph of 'Theoretical Explanation', that explanations are not arguments, but I do think it would be salutary for the whole discussion of scientific explanation if this alternative could at least be seriously entertained.

One can draw an analogy—but it is only an analogy, and should not be misinterpreted as an identity—between this issue of the relation between explanation and argument on the one hand and the issue, raised by Carnap, of the relation between confirmation and inductive inference on the other. As is well known, Carnap denied that there are, strictly speaking, inductive arguments; inductive logic, he maintained, contains no rules of acceptance. This was a highly counter-intuitive claim, it had profound consequences, and it has been intensely debated in the literature. Regardless of which side of this controversy one ends up on, there is no doubt that the issue is substantive and important, and that our understanding of inductive logic has been significantly deepened as a result of the debate. A similar benefit, it seems to me, may well result from a radical questioning of the thesis that scientific explanations are arguments.

Second, one way of trying to get at the question of what deserves to be called an 'explanation' is to consider what sort of value explanations might have. In his famous article, 'Deductive-Nomological vs. Statistical Explanation', Hempel tried to deal with this question, but without any very satisfying result.[4] Greeno, in contrast, offers an excellent answer to this very question as applied to the statistical-relevance model of explanation: the value of an S–R explanation can be measured

[4] In Herbert Feigl and Grover Maxwell, eds., *Minnesota Studies in the Philosophy of Science*, vol. III (Minneapolis: University of Minnesota Press, 1962), esp. pp. 153–6.

directly by the way it increases our information—by the *information transmitted*, in the technical sense of that term.[5] In defining 'degree of inhomogeneity' of a reference class, I offered essentially the same measure of the value of an *S–R* explanation, framed in terms of predictive utility.[6] (Popperians, who believe that science has no predictive import, will naturally be unmoved by this consideration.) The fact that *S–R* 'explanations' increase our information in a precisely measurable and specifiable way lends some rationale to identifying them with what we would have taken pre-analytically to be scientific explanations.

In a more practical vein, I attempted to answer the question in the following terms:

> One might ask on what grounds we can claim to have characterized explanation. The answer is this. When an [*S–R*] explanation . . . has been provided, we know exactly how to regard any *A* with respect to the property *B*. We know which ones to bet on, which to bet against, and at what odds. We know precisely what degree of expectation is rational. We know how to face uncertainty about an *A*'s being a *B* in the most reasonable, practical, and efficient way. We know every factor that is relevant to an *A* having the property *B*. We know exactly the weight that should have been attached to the prediction that this *A* will be a *B*. We know all of the regularities (universal or statistical) that are relevant to our original question. What more could one ask of an explanation?[7]

These remarks pertain to all types of *S–R* explanations, even those that in a particular case lower or leave unchanged the probability of a particular explanandum-event.

Still, I do not want to engage in a battle for a word. My non-verbal claim can be put as follows: what I have called

---

[5] 'Explanation and Information', reprinted in *Statistical Explanation and Statistical Relevance*.
[6] *Statistical Explanation and Statistical Relevance*, pp. 51–3.
[7] Ibid., p. 78.

'statistical-relevance explanation' does the job Hempel's so-called 'inductive-statistical explanation' was designed to do, and, I have tried to argue, the *S–R* model does it better. Whether correct or incorrect, this thesis in neither verbal nor trivial.

2. Cohen's second point centres upon the perfectly accurate observation that many of my claims about explanation depend, not just upon definitions of such concepts as the screening off relation, but also upon features of currently accepted physical theories. I made no bones about this fact. In footnote 32, which refers to the claim that coincidences are to be explained by common causes in the intersections of the backward portions of the Minkowski light cones, I said,

> These statements obviously represent factual claims about this world. We believe they are true, and if they are true they are very important. But we have no reason to think they are true in all possible worlds.

And at the end of my discussion of spatio-temporal continuity of causal relations, I added,

> A *complete* set of factors statistically relevant to the occurrence of a given event can be found by examining the interior and boundary of an appropriate neighbouring section of its past light cone. Any factor outside of that portion of the cone that is, by itself, statistically relevant to the occurrence of the event in question is screened off by events within that neighbouring portion of the light cone. These are strong factual claims; if correct, they have an enormous bearing upon our conception of explanation.

Reichenbach was certainly emphatic in maintaining that his *principle of the common cause* represents a physical fact about this universe,[8] and Russell, in *Human Knowledge*, seems totally committed to the view that inference and understanding

[8] Hans Reichenbach, *The Direction of Time* (Berkeley and Los Angeles: University of California Press, 1956), §19.

depend upon physical postulates.[9] In *Statistical Explanation and Statistical Relevance* I explicitly affirmed the factual foundation for my conception of statistical explanation.[10]

The question Cohen raises is whether this feature of my approach constitutes a defect. '[P]erhaps . . . we need to distinguish between those constraints on scientific explanation which are very deeply rooted and are usefully regarded as belonging to the logical structure of explanation and those constraints, like the exclusion of action at a distance, which are somewhat less deeply rooted and may better be conceived as depending on currently accepted physical theory.' Now I would be the last to deny the importance of such a distinction, though I think it is essential to be quite clear on how the term 'logical' is to be construed. *If* one adopts something like the deductive-nomological model of explanation, it is useful to recognize that deductive validity, and perhaps even lawlikeness, are logical concepts in some reasonably strict sense. But if one goes on to maintain that temporal priority is part of the 'logic of the term "cause" ' or that it is part of the 'logic of the explanation concept' that the events invoked to explain a given event must precede it, I fear we are courting serious confusion. For 'logic' in this sense seems to be little more than the common sense physics that has become enshrined in ordinary language. If we are going to appeal to physics, I agree fully with Cohen that we should be explicit about it, and, it should be added, we might as well use the best physical theory available rather than the worst.

But once all of this has been acknowledged, the question is how far we can go in characterizing scientific explanation in any interesting way if we stick to the logical machinery alone, without invoking fundamental physical theory. The answer, I think, is *not very far*. For example, temporal asymmetry seems to me one of the most basic features of scientific

[9] Bertrand Russell, *Human Knowledge, Its Scope and Limits* (New York: Simon and Schuster, 1948), Part VI.
[10] pp. 75–6.

explanation. I consider the failure of the deductive-nomological account to embody that feature one of its chief defects. Acceptance of the consequence that an eclipse can as well be explained in terms of subsequent configurations of the sun, moon, and earth as in terms of prior configurations must be a leading instance of the principle that one man's counterexample is another man's *modus ponens*. However, as has been convincingly argued by Reichenbach, Grünbaum, et al., the temporal asymmetry or anisotropy of the world is a matter of fact rather than a matter of logic. Reichenbach was very explicit in pointing out that the principle of the common cause is contingent; I follow him in insisting upon that point, even though I think it may be the most fundamental principle of scientific explanation. I regard it as crucially important to our concept of scientific explanation that the Minkowski light cone *is* the cone of causal relevance, but that surely is a nomological matter rather than one of logic. And the principle of action by contact which seems to work well in the macrocosm, but not in the microcosm, is obviously no kind of logical principle. Indeed, it seems to me a most challenging problem to devise a concept of scientific explanation which is suitable for the quantum domain. Logic, I fear, will not be of much help except in a negative way—except, that is, to keep us from such blunders as the view that explanation is impossible with regard to events that are not causally determined. Hence, it seems to me, though we do need to acknowledge the distinction between logical principles and physical principles invoked in our attempt to understand explanation, we certainly have to cross that boundary to have much hope of arriving at anything like an adequate characterization of scientific explanation.

In the final paragraph of his remarks, Cohen claims that my discussion of causal explanation 'merely urges a supplementation of Hempel's analysis' and 'offers a partial elucidation of this supplement in terms of screening off'. This may be all I succeeded in doing, but it is certainly a far cry from what I was attempting to do. My aim was to provide (at least a prolegomena to) an elaboration of a radical alternative to the standard D–N model of explanation. It began, on the lowest level so to speak, with a summary of the statistical-relevance model of explanation which differs from the deductive-nomological or inductive-statistical models in no less fundamental a fashion than *to deny that explanations are arguments.* The main purpose of the present paper was to extend the statistical relevance model to the higher level of causal/theoretical explanation—a level on which statistical generalizations rather than particular events are to be explained. It remains a fundamental principle that explanations are *not* arguments. A crucial part of the attempt is to try to understand the involvement of causal relations in explanations—a matter rather casually dismissed in many treatments of the deductive-nomological model. I do not believe it should be so summarily dispensed with; it seems to me that there is an important intuition underlying the feeling that Kepler's laws (even if generalized to eliminate explicit reference to our Sun) can at best provide a very superficial explanation of planetary motion, while Newton's laws can supply a much deeper explanation. It seems to me, moreover, that the greater generality of Newton's laws does not entirely account for the difference, but that the main characteristic is the causal (dynamical) nature of Newton's laws. I am painfully aware of the fact that much more needs to be done to provide anything like a complete treatment of this problem, and a complete defence of the alternative model. But the present paper represents a start.

According to the account of causal/theoretical explanation offered in my paper, a causal explanation of a statistical regularity is a description of how that statistical regularity arises out of causal connections among events (or types of events) in the world. Such an explanation is not an argument, so *a fortiori,* it is not a Hempelian explanation with some added constraints. It is, instead, more like a model of the universe in which the causal nexus is displayed. Whether such an account can be successfully elaborated I cannot say; I can only urge that it be looked upon as an attempt to devise a model of explanation that differs drastically from the standard D-N account, not as an effort to provide needed refinements of the received view.

2 REPLY TO DR. MELLOR

Among the three objections raised by Dr. Mellor, it seems to me that the first two involve serious issues, while the third rests upon simple misunderstanding. Since the third objection, if well-founded, would be devastating, I shall deal with it first. I shall then discuss his second objection, and finally his first, thus reversing the order in which he presented them.

1. Mellor accuses me of conducting 'a circular tour of uncomfortably small radius', but I fear that what appeared circular to him was actually elliptical. Epistemologists have often reminded us that a circle, viewed from certain angles, may appear to be an ellipse, but they have not so often warned that an ellipsis, seen from the wrong angle, may look like a circle. Three paragraphs above the sentence Mellor quotes in part, 'Assuming . . . that a temporal direction has been established, we can say . . .', I gave what must be regarded as at least a broad hint as to the basis for this assumption:

> Since we are not, in this context, attempting to deal with the problem of 'time's arrow', and correlatively, with the nature and existence of irreversible processes, let us assume that we have provided an adequate physical basis for identifying irreversible processes and ascertaining their temporal direction.

This statement occurred in the context of a discussion of Reichenbach's 'mark method' for distinguishing causal processes from pseudo-processes, and it was accompanied by references to Reichenbach's writings on this subject. Indeed, the very statement quoted by Mellor is annotated with the following remark:

Although Reichenbach seemed to maintain in his earlier writings, such as *The Philosophy of Space and Time*, that the mark method could be taken as an independent criterion of temporal direction (without any other basis for distinguishing irreversible processes), he abandoned that view in the later work, *The Direction of Time*.

I can only add that this posthumous work of Reichenbach was, as the title suggests, devoted primarily to precisely this problem of whether the *later than* relation is to be explicated in terms of causal asymmetries, or conversely, His unequivocal answer is that the causal asymmetry is the more basic one. In Chapter IV of this work he offers a detailed analysis in terms of the principle of the common cause, a principle upon which I relied heavily in my paper. But those who, like Mellor, doubt the efficacy of this approach can be referred to the detailed microstatistical analysis of temporal asymmetry provided in Chapter III of the same work. This will serve equally well to support my assumption that we have an independent, non-question-begging, non-circular criterion of temporal asymmetry based upon irreversible processes (in the thermodynamic sense). To keep the length of the original paper within reasonable bounds, I did not spell this out in detail, but perhaps a brief word of explanation is in order here. Reichenbach's basic strategy is to construct what he calls a 'causal net' on the basis of causal connections which are symmetric (as in the laws of classical mechanics) or which, if asymmetric, are used without exploiting their asymmetric character. Once this causal net has been constructed, further considerations can be adduced to endow it with a causal or temporal asymmetry. For example, suppose a billiard ball travels between points A and B. This statement is temporally neutral; we are not specifying whether the direction of travel is from A to B or from B to A. Nearby, we have a container with two compartments with an opening between. When the ball is at B, we note that one compartment is full of oxygen, while the other compartment is nearly empty.

When the ball is at A, we note that both compartments contain oxygen at nearly equal densities. We say, then, that the ball travelled from B to A—i.e. it was at B earlier than it was at A. On the basis of thermodynamic considerations, we judge this kind of diffusion process to be irreversible.

Now, the reason that the mark method, by itself, is inadequate to establish an unambiguous temporal direction is, essentially, that we have to understand the distinction between imposing a mark and removing a mark—for example the difference between making a chalk mark on a billiard ball and erasing such a mark. But this difficulty does not mean that we cannot use the mark method for establishing causal connections, or tracing out causal processes, provided we remain non-committal regarding their temporal directions. Suppose, for example, that there is a beam of light shining on a wall, and we notice that the colour of the spot on the wall changes from white to red and red to white from time to time. Suppose that we also notice a red filter which is moving in and out of the beam. Suppose, moreover, that the presence or absence of the filter in the beam is highly correlated with the colour of the spot on the wall. We have said nothing, as yet, about the filter causing the beam to be red; so far, it is equally open to say that the presence of the red filter is caused by the colour of the spot on the wall. Nevertheless, we can say, without committing ourselves to a temporal direction, that there is a causal process present here. With the aid of such symmetrical causal connections, we can construct the kind of causal net Reichenbach proposed. The net is then examined for 'causal forks'—i.e. situations in which we can say, still without introducing any temporal asymmetry, that we have a pair of events with either a common cause or else a common effect (but not both). We *then* invoke the principle of the common cause to determine that it is a common cause rather than a common effect. The causal or temporal asymmetry is thus introduced. This asymmetry can then be exploited to ascertain the directions of the various marking processes that have been used to construct the causal net. Some marking processes, such as light absorption by a filter, are then seen to involve irreversible processes, and these can then be used to distinguish between cause and effect.

I have, of course, given only the briefest sketch of the two ways of introducing the asymmetry upon which the asymmetry of the relation of causal relevance or causal influence is based. The details can be found in Reichenbach's book. But what I have said will, I hope, at least give some indication of the way in which I believe Mellor's charge of circularity can be circumvented.

2. Mellor's second difficulty concerns my 'use of "marking" as a test for causal processes. Marking seems to me merely a special case of causation, not a more primitive notion in terms of which causation can usefully be explicated.' He summarizes his discussion of this point by saying, 'I find that marking, as a merely special case of causation, cannot play the rôle in its explication that Salmon assigns to it.'

Now, it is impossible to deny the correctness and importance of Mellor's basic observation: marking is, indeed, a special case of a causal *interaction*. This does not mean, however, that marking has no value in explicating causal *processes*—i.e. propagation of causal influences across spatio-temporal intervals. In my paper, I was mainly concerned with the explanation of non-local statistical correlations, and I discussed such explanations in terms of connections of the correlated events via spatio-temporally continuous causal *processes*. I was remiss in not making this distinction explicitly and emphasizing its importance; however, it is straightforward and reasonable, and it obviates the problem Mellor rightly pointed out.

By repeated use of such devices as Mill's methods of agreement and difference, we may establish that pulling the trigger of a loaded gun results in a bang. This is a local coincidence (and, as our knowledge increases, we can explain this interaction in terms of causal processes and more highly localized interactions). To establish a connection between such events and the demise of a person at a different space-time location, we provide a causal process, in the form of the motion of a bullet, connecting them. Moreover, as any reader of detective fiction knows, the bullet carries marks that enable one to identify the gun from which it came. The basic problem is this. Given knowledge of local causal interactions, how can we ascertain the causal structure of the world? This causal structure is, of course, intimately related to its space-time structure.

In terms of what we know at present, the Minkowski light-cone plays a fundamental role in that structure; indeed, as I claim in my paper, it defines the limits of causal relevance. Given two events, it is an invariant fact of nature, not relative to any particular frame of reference or vantage point, that the space-time separation between them is space-like, light-like, or time-like. Which type of separation obtains is determined by the physical possibility of connecting them by various types of causal process: it is space-like if they cannot be connected by any type of causal process, it is light-like if they can be connected only by causal processes that qualify as 'first signals', and it is time-like if they can be connected by other types of causal processes as well.[11] These facts are, I realize, very familiar, but their fundamentality needs to be emphasized, so I beg indulgence to reiterate them.

If two events exhibit a space-like separation, they may be pronounced simultaneous, or either may be taken as earlier than the other, depending upon which frame of reference is adopted for describing them. However, they may be connected by some sort of 'process'—what Reichenbach called 'unreal sequences' and what I called 'pseudo-processes'—such as the moving spot of light on the distant clouds or the scanning pattern on a TV tube.[12] The distinction between causal

[11] A 'first signal' is one whose rate of propagation is maximal, and hence, is equal to the speed of light.

[12] Oscilloscopes are commercially available with a 'writing speed' of $4/5$ c, and no reason in principle prevents them from achieving writing speeds greater than c. If pulsars are, indeed, rapidly rotating sources of radio signals, they actually exhibit the type of pseudo-process involved in the 'thought experiment' with the rotating beacon shining upon distant clouds.

processes and pseudo-processes must be fundamental, for if pseudo-processes could be used to synchronize spatially separated clocks, the relativity of simultaneity would vanish, and special relativity would be destroyed by re-establishment of relations of absolute simultaneity. We must ask, therefore, what basic feature distinguishes causal processes from pseudo-processes. The most acceptable answer seems to be that causal processes can be used for purposes of signalling or transmitting information—i.e. for the transmission of marks. The practical utility Mellor acknowledges rests, I continue to believe, upon a fundamental feature. The importance of the distinction between causal processes and pseudo-processes cannot be denied. If this distinction does not rest upon the characteristic of mark transmission, then there is serious need to discover in what it does consist.

When Mellor claims that 'Marking is in principle redundant; if we cannot detect causal processes without marking, then in particular we cannot detect marking', he seems to me to be overlooking the importance of pseudo-processes. This defect, incidentally, also mars Russell's otherwise illuminating views on causal processes. Bearing in mind the acknowledged distinction between local interaction and causal propagation, we must distinguish among the following sorts of cases: (1) A beam of white light travelling from A, past B, to C, (2) A beam of red light sent from A, via B, to C, and (3) A beam of white light sent from A to B, where it passes through a red filter, and continues on from B to C as a beam of red light. Alternative (1), even if frequently repeated, does not establish the process of light travelling from A to C as a causal process, for the spot of light cast by the rotating beacon on the clouds also makes its trip repeatedly, without thereby qualifying as anything other than a pseudo-process. Alternative (2) is no help, for by fitting the rotating beacon with a red glass, we can easily achieve the repeated transit of a red spot as another pseudo-process. Alternative (3) is thus crucial. In order to establish a process as genuinely causal, we must be able to show that a *local* interaction at some point in that process can produce a modification of structure which will be transmitted from that point on through the remainder of the process (or at least a significant subsequent portion of it).[13]

Mellor does identify one problem connected with the criterion of mark transmission; namely, that the mark imposed at one place may reappear in another without a genuine transmission from the one to the other. This might happen in either of two ways. First, it might occur completely fortuitously. Suppose we make the spot on the cloud red at one point by interposing a red filter in the light beam, but we do not move the filter with the beacon, so we do not continue to mark the processes that are genuinely causal.[14] We look at the moving spot of light on the cloud at another place; lo and behold it is red (because a piece of red cellophane chanced to blow past the beacon just as it was casting light upon the spot we happened to be examining). Such coincidences must be admitted as occasional possibilities, but they are to be weeded out by repetition of the experiment. We may also observe the light spot at intermediate points; if a genuine transmission of the mark occurs it will be a continuous transmission through the intervening points. That such a coincidence *might* happen again in precisely the places we looked just as we were re-peating the experiment is merely an instance of the essential fallibility of scientific knowledge. We need worry about such coincidences even less than we do about our assumption that two identical essays have a common source.

A second possibility is the existence of some more or less elaborate triggering mechanism that makes a red light appear at

[13] A mark can, of course, be 'erased' in the natural course of events. A chalk mark on a billiard ball, for example, will gradually disappear as the ball rolls around on the table.

[14] If the filter is placed in the beam very close to the cloud, thus providing a local interaction, it will be physically impossible (because, by hypothesis, the spot of light moves with a speed greater than light) to move the filter in such a way as to keep it within the beam of the rotating beacon, and thus effectively to continue marking the pseudo-process.

one place shortly after a light beam has been made red at another place by interposition of a filter. This is accomplished without the aid of a beam of red light which traverses the entire interval between these two occurrences of red light. For example, a man with a flashlight stands atop a tower, and at a particular moment, he covers his light with a red filter Another man, stationed some distance away, sees the red light; when he does, he picks up his telephone, calls a third man, and tells him that he has just seen the red light. The third man stands beside a console of buttons which enable him to light lights of various colours. When he receives the call from the second man, he pushes the button that turns on the red light. Thus, a red light has occurred at the end of a causal process without having been transmitted as a mark. The point, I suppose, is that the man at the console could have turned on a light of any colour, though he happened to choose red, while a beam of light, rendered red by a filter, cannot spontaneously turn green.

The key to one answer to Mellor's objection resides in the term 'spontaneously'. Regarded as a closed or quasi-closed process, a light ray will retain its character either as a random mixture of frequencies or as a monochromatic ensemble. To say that a light beam is a closed or quasi-closed process means simply that it does not suffer local interactions with its immediate surroundings. I have acknowledged the importance of the distinction between local causal interactions and causal processes; it can be utilized to characterize closed and quasi-closed systems. When we are dealing with marks which involve some kind of sorting, an imposition of structure, or a modification of structure, we are treating situations in which a low entropy state is produced as a result of a local interaction with the environment. If a mark is produced at one point by sorting the red frequencies from all the other frequencies and this mark is *not* transmitted to another place, the same type of sorting must be done by a similar kind of local interaction if the red colour is to reappear. Thus, in characterizing transmission of a mark from one place to another in a process or pseudo-process, we may decide to exclude the existence of another sorting interaction of the same type elsewhere in the process (or pseudo-process). This is not a question-begging restriction to impose.

There is, however, an alternative way to look at the same situation. What is important in the transmission of a mark is not its qualitative features, but its structural character. Structure, which represents order, can be identified with low entropy and with information. Ability to transmit information is the crucial feature of causal processes, and qualitative homogeneity of the process is of secondary importance (if it is of any importance at all). Russell discussed this point at length, and provided many examples.[15] For instance, a vocalist sings a song. He reads symbols on a sheet of paper which convey to him the melody and the words. By complex processes involving the transmission of light, neural responses, transmission of neural impulses, muscular contractions, etc., he transforms a structure represented by the symbols into a complex pattern of sound waves that are transmitted to a microphone. The microphone, n turn, translates this pattern into a complex pattern of electric impulses which are then transformed into mechanical changes in a recording machine. The mechanical impulses produce patterns in grooves on a plastic disk, and these are reproduced on other plastic disks. You purchase one of these other plastic disks, carry it to your house and place it on your hi-fi. By another familiar, but relatively complex process, it translates these patterns back into the sound of the singer's voice. Such transmission of structure and information, through a long series of physically heterogeneous processes, can be achieved only by means of causal processes. Pseudo-processes would be unavailing, and that is what is crucial. The test of whether a spatio-temporally continuous series of similar structures can transmit information is to see whether a modification of that structure (for example the singer sings one note off-key)

[15] See, for example, *The Analysis of Matter* (New York: Dover Publications, 1954), ch. XXVI.

reappears throughout the series without additional local interventions of an analogous sort to reproduce the same kind of modification. For example, a 'singer' who pretends to be singing, but is actually only mouthing while a recording is being played behind a curtain, is inevitably given away if the needle fails to 'track' correctly.

3. I come finally to Mellor's first objection, which is, as he says, 'a general objection to any explication of causation in terms of statistical relevance as understood by Salmon'. His argument on this point involves a number of aspects; I shall try to sort them out.

Let me begin by taking up what seems to be offered as *the coup de grâce* of the entire statistical approach to causality: 'Yet if the literature of causation agrees on anything, it is surely that e can only cause e' if both e and e' at least *occur*.' This statement is unexceptionable, but its apparent relevance to the issue at hand is completely spurious. Almost everyone would agree that decapitation causes death, and that in particular, Lavoisier's death was caused by the guillotine. None of this is in any way incompatible with the recognition that Hume's death was not caused by decapitation. Even in the most deterministic of circles, it has been widely acknowledged that there can be alternative kinds of causes for a particular sort of effect. In acknowledging that events of type E cause events of type E', we are not committed to saying that an event of type E' occurs if and only if an event of type E occurs. And the statement that e (Lavoisier's decapitation) causes e' (Lavoisier's death) only if e and e' both occur (where e is of type E and e' is of type E') does not entail that an event of type E' (death) occurs if and only if an event of type E (decapitation) occurs.

Once this point is clearly understood, the question whether causal relations obtain between particular events or between classes of events is seen to lose its apparent significance in the present context, no matter how interesting an issue it might be in its own right. The question of whether one particular event can properly be considered the cause of another must hinge, as

Mellor acknowledges, upon the properties of these events. This issue was raised, as nearly as I can tell, to lend added force to the point on which 'the literature of causation agrees'.

Much the same can be said for the charge of inconsistency Mellor seems to level at the 'merely statistical analysis'. Surely no one would want to assert both, 'If the transmission had not occurred, other things being equal, the pattern would not have occurred,' and also, 'On a particular occasion on which other things *were* equal the pattern occurred without the transmission.' But when we recognize that Hume's death from other causes is compatible with Lavoisier's death being caused by decapitation, we see that the first statement need not be taken—Mellor to the contrary notwithstanding—as a *sine qua non* of any analysis of causal relations. I am fully aware, of course, that other things were not equal in the deaths of Hume and Lavoisier. The point is merely that we can, and do, properly say that an event of type E caused an event of type E' without being committed to saying that no event of type E' can occur unless it is caused by an event of type E. The *ceteris paribus* clause raises further issues which I shall take up below.

In view of these considerations, we see that it is entirely reasonable to say that a particular crash was caused by a blow-out even though not every crash is caused by a blow-out and not every blow-out results in a crash. Thus, the point on which Mellor finds the entire 'literature of causation' in complete agreement has no bearing whatever upon the question of whether causality can be explicated in fundamentally statistical terms. Moreover, it is obvious that the 'literature of causation' does *not* agree on the non-statistical character of the causal relation—witness, for example, works of I. J. Good, Patrick Suppes, and Hans Reichenbach.[16] And this is not because these

[16] I. J. Good, 'A Causal Calculus', *British Journal for the Philosophy of Science*, vol. 11 (1961), and vol. 12 (1962); Patrick Suppes, *A Probabilistic Theory of Causality* (Amsterdam: North-Holland Publishing Co., 1970); and Hans Reichenbach, *The Dircetion of Time* (Berkeley and Los Angeles: University of California Press, 1956).

authors have failed to see the implications of the point on which the 'literature of causation' does agree.

Mellor, of course, acknowledges this point, for he states explicitly that he is '. . . not claiming that a 100 per cent correlation is necessary for a singular cause-effect relation to obtain . . .'. What he is claiming, if I understand him rightly, is that in those instances in which an event of type E occurs, but is not followed by an event of type E', that particular member of E had some additional relevant feature which accounts for the fact that it did not have a member of E' as its effect. This is quite plausible for most of the everyday examples we have mentioned. We can, in principle, find additional factors to explain why this blow-out did not lead to an accident (the car was not travelling very fast, there was not much traffic, the shoulder was wide and smooth), or why this radio transmission did not cause breakup of the TV picture (atmospheric conditions, wrong frequency, set turned off).

But consider some slightly more problematic examples. Suppose that metal balls one centimetre in diameter are dropped vertically on to a very sharp vertical knife-edge. Some balls will surely bounce off and come down in a box on the right side, and others will surely bounce into a box on the left. The language and concepts of causality seem appropriate in such cases. But still other balls will fall with their centres almost directly above the edge; they will bounce upward and come down once more, striking the knife-edge again. Eventually, since this sort of equilibrium is highly unstable, each ball will fall to one side or the other. Are there, even in principle, relevant factors that determine for each ball whether it will eventually end up on the left or the right? I think it would be physically unsound to claim that such relevant factors must exist in each case; to say the contrary would seem to me a dogmatic and unfounded version of traditional determinism.

I do not believe that Mellor wishes to commit himself to this sort of a priorism; that is, he does not seem to be asserting that every event is causally determined down to its last detail. It seems that he is arguing, rather, that the language of causality is to be applied only in those cases where full causal determination obtains, and withheld in those cases in which there is any degree of indeterminanacy. But if this is his view, I wonder how he would treat the example of the balls falling upon the knife-edge. If the language of causality is appropriate for some of the balls that bounce to one side or the other, but not for others, where is the line to be drawn between cases of the two types?

Mellor's approach seems, actually, to involve difficulties even more severe, for he apparently holds that causes are necessary as well as sufficient; 'If the transmission had not occurred, other things being equal, the pattern would not have occurred,' is offered as a reasonable requirement for causation to obtain in that instance. Suppose we attempt to impose the same requirement upon the balls bouncing off of the knife-edge. If we wish to say that the *cause* of a ball bouncing to the right was the fact that its centre was three millimetres to the right of the knife-edge upon impact, we must be prepared to say that it satisfied certain conditions which are necessary for it to fall to the right. I am assuming that striking the knife-edge with its centre three millimetres to the right is sufficient. Certainly we could not maintain that striking the knife-edge with centre exactly three centimetres to the right is necessary as well as sufficient, for it would be sufficient if the centre were four millimetres to the right, and (let us assume as well) it would be sufficient if the centre were two millimetres to the right. But we must be able to say, for a small distance (say a few microns if the experiment is being done with precision), that the ball will bounce to the right if *and only if* the centre of the ball is at least that distance to the right of the knife-edge upon impact. This requirement would obviously rule out any area of indeterminacy in which it is an irreducibly statistical matter which way the ball bounces. Such a requirement seems utterly excessive. In order to say that the *cause* of a particular ball bouncing to the right was the fact that its centre was three millimetres to the right of the knife-edge, we must be prepared to say that it is possible in principle to ascertain with unlimited precision the conditions that are necessary and sufficient for *any* ball to end up in the right-hand box. The only escape I can see from this consequence is to maintain that we can, in principle, ascertain with unlimited precision from the subsequent state of the ball whether it got where it is by a fully deterministic causal process or by a non-deterministic stochastic process. This claim seems even harder to swallow than crude Laplacian determinism. These are the sorts of problems hidden under Mellor's *ceteris paribus* clause.

Let us consider another common-sense example. Waves of a certain size are large enough to capsize a particular toy boat, and waves of another size are too small to make it capsize. Unless we insist upon determinism, we shall have to say that in between are waves within some small range with which we can at best associate a probability distribution for overturning the boat. Within this range no relevant factors can be summoned, even in principle, to determine with certainty the cases in which the boat will overturn, and in which it will not. If a particular stone is dropped into the pond. where the boat is sailing, it will set up a wave front that moves out concentrically, but whose size (amplitude) diminishes with increasing distance from the centre. Thus, depending upon the relative positions of the boat and the place at which the stone enters the water, there will be some cases in which the boat is almost sure to capsize, others in which it will very probably not founder, and some in which the chances are somewhere in between. This situation seems to typify many important causal processes. A given causal 'influence' is generated at some space-time location, and it diminishes as it spreads to more and more distant regions. It seems natural to describe this attenuation in statistical terms.

The example of the toy boat is similar in important respects to the example of the balls bouncing off of the knife-edge. In cases which involve various forms of unstable equilibrium it is hazardous to claim the existence of deterministic causes. Yet, the language of causation seems appropriate enough. Indeed, even in cases acknowledged to be out-and-out statistical, we are willing to talk in terms of causal relations if the probabilities are high enough.[17] We say, for instance, that heating the gas in a closed container of fixed size causes the pressure to increase, even though we admit that there is a minute but non-vanishing probability that the added kinetic energy will involve motion in a single direction rather than randomly distributed directions. In that case the net effect of heating the gas would be to make all of the molecules rush towards one side of the container, thus causing it to move off (say) to the left. We countenance causal terminology in such contexts because the probability of the abnormal result is so low as to 'make no odds', and consequently the probability of the normal outcome is extremely high. The question then becomes, how high is high enough? Those who have followed the controversy over acceptance rules in inductive logic will recognize this as a deeply embarrassing question.

At this point we may come under suspicion of involvement in another merely terminological dispute, this time regarding the applicability of the term 'cause' in non-deterministic contexts. The natural suggestion would then be to adjudicate the dispute by agreeing (as might seem best to accord with ordinary usage) to reserve causal language for deterministic contexts, and to speak of statistical relevance rather than causal relevance in non-deterministic settings. This simple solution will not do, however, since it would completely obscure the fact that the kinds of relevance relations I have called 'causal relevance relations' have many important properties not possessed by other kinds of statistical relevance relations. The main purpose of my paper was to explicate these relations and examine their rôle in explanatory contexts. To revert to calling them simply 'relations of statistical relevance' would seriously cloud the

[17] See Richard Jeffrey, 'Statistical Explanation vs. Statistical Inference'.

major issue. I could coin a new term, for example, 'statisticausal relations', but I do not feel obliged to do so, since the term 'causal relevance' still seems quite suitable for my purposes. Artificial terminology, like other sorts of entities, should not be multiplied beyond necessity.

IV/Ideological Explanation

J. L. Mackie

An ideology is a system of concepts, beliefs, and values which is characteristic of some social class (or perhaps of some other social group, perhaps even of a whole society), and in terms of which the members of that class (etc.) see and understand their own position in and relation to their social environment and the world as a whole, and explain, evaluate, and justify their actions, and especially the activities and policies characteristic of their class (etc.). Thinking in terms of this system unites and strengthens that class and helps to maintain it and to advance its interests. This system is determined by the social existence of the class (etc.) of which it is characteristic. It is not in general deliberately invented or adopted (though it may be deliberately fostered and propagated). At least some of the beliefs and concepts in the system are false, distorted, or slanted, and at least some of the activities sustained and guided by the ideology have a real function different from that which, in the ideology, they are seen as having.

This notion of an ideology is, I think, fairly widely employed. It is due primarily to Marx, but similar notions, sometimes under different names, have been developed by Pareto, Sorel, and Mannheim. But I shall not be concerned with scholarly questions about the ascription of such a notion to this or that thinker. Nor am I primarily concerned with the empirical question how widely applicable or how useful the notion is. Rather I shall discuss certain difficulties and problems that the notion involves or might be thought to involve and consider what bearing the notion has upon the question, 'What kinds of explanation are possible in the social field?'

The phrase 'ideological explanation' which I have taken as my title is, of course, radically ambiguous. (1) An ideological explanation might be an explanation given within and in terms of an ideology by some of those whose ideology it is—for example, an explanation of inflation given by bourgeois economists using concepts and beliefs that belong distinctively to the bourgeois ideology. (2) An ideological explanation might be one given by someone who refers, as from the outside, to an ideology. We could further distinguish, within this, between (*a*) an explanation of the ideology itself or of some part of it—perhaps showing how, in Marx's famous phrase, it is determined by social existence—and (*b*) an explanation of something else that occurs, for example some actions or behaviour of some of those who have a certain ideology (or of some further social phenomena to which such actions contribute) by reference to the fact that they have this ideology.

These three varieties of ideological explanation give rise to three main groups of problems.

(i) Must an ideology be false or distorted? If this feature is included—as I have included it above—in the definition of ideology, we can still ask whether this feature necessarily accompanies the other features that have been mentioned in the definition. And if an ideology is false or distorted (whether or not it is necessarily so), must this vitiate any explanation of kind (1), that is, any explanation given within the ideology? An associated question is whether it can be coherently maintained that all thinking about society is ideological, in particular whether the doctrine of the general 'sociology of knowledge' is defensible or self-refuting.

(ii) One problem in the first group was whether explanation of kind (1) is radically faulty because it is irretrievably biased. A similar problem in the second group is whether explanation of kind (2*b*) is radically faulty because it explains something other than what it purports to explain. It purports to explain actions and social behaviour: but the ideology of the relevant class is a constituent of those actions and behaviour—they are essentially however the agents see them to be—so any explanation from outside, which treats the ideology merely as a fact, causally related to actions and behaviour described in other terms, will inevitably neglect something essential. This point and the previous one together set up a dilemma: if you try to explain social phenomena from inside the relevant ideology, your explanation is hopelessly biased and distorted, but if you try to explain them from outside the ideology you are ignoring something essential to those phenomena. If this dilemma can be evaded or overcome, there is the further question whether any explanation of actions by ideology can be a causal one, or whether actions and ideology are too intimately related to be such distinct existences as cause and effect, at least on anything like a Humean view of causation, are required to be. This is obviously a variant of the old issue about reasons and causes.

(iii) There is a similar problem about explanation of kind (2*a*): just how can social existence determine consciousness? If this is supposed to be a causal relation, can we find a 'social existence' which is logically separate enough from the ideology to cause it? Does not anything that we can call social existence already incorporate the associated ideology? Also, is it at all plausible to suggest that this is a case of one-way causation, that social existence determines consciousness whereas consciousness does not determine social existence, as Marx and Marxists have frequently maintained—though they have also frequently maintained the opposite, that mental and material factors interact. Are ideologies epiphenomenal or not?

Taking this last question first, I would insist that any strict doctrine of one-way causation would be utterly implausible. As Popper points out (*The Open Society and its Enemies*, vol. II,

p. 108) the changes made in Russia by Lenin and his successors are a clear example of ideas revolutionizing economic conditions. And in any case we can understand the evolution of ideologies only on the assumption that they perform some function—in my original definition of an ideology I said that it unites and strengthens the class and helps to maintain it and to advance its interests. An ideology would have no survival value if it did not react on social existence at least in these ways. If Marx's thesis is to have any truth or even plausibility it must be taken in some different sense. It could be taken first as a denial of one-way causation in the other direction, as saying that ideas do not spring either from nowhere or only from other ideas and then control social policies and social structure, but are themselves affected by the other, for example economic, aspects of society. It could be taken secondly as saying that aspirations and schemes for social betterment are powerless if other conditions are not ripe for their implementation, for example as condemning 'utopian socialism'. Both these points are now truisms, but perhaps only because Marx has made them so. But a third and now more interesting interpretation of Marx's dictum would be this. Although 'social existence' and 'consciousness' interact, we can still ask in which of them major changes typically originate, whether the material or the mental is the leading edge of social change, and we might answer that it is the material. It is, however, very doubtful whether this answer can be true. We must, of course, distinguish as Marx did between the forces of production—that is, resources and technology—and the relations of production—that is, economic structures, as well as contrasting these two together with political and legal superstructure and consciousness or ideology. But then it seems plausible to say that sometimes new resources and technology initiate social change, sometimes new relations of production do so, and at least sometimes ideology takes the lead, particularly where economic structures that already exist in some countries are introduced into others from which they have been absent—Russia after Lenin, as mentioned above, is a case in point. We have a genuine form of question here, but not one that admits of a universal answer.

The other problem in this group was whether we can find a 'social existence' which is sufficiently distinct from consciousness or ideology to be even a possible cause of the latter. (This question seems to worry Plamenatz in Chapters 2 and 3 of *Ideology*.) Now of course there never is or can be a social existence as a concrete whole, an actual collection of goings on, which does not involve ideas and ideology. Concrete social existence always involves, among other things, conventions, established patterns of behaviour supported by various sorts of pressure. Any concrete relations of production will thus have an ideological aspect or component. Equally resources are resources only in so far as they are recognized as being such—or, more strictly, potential resources become actual resources only when they are recognized as being potential resources—and technology obviously includes an intellectual component, both knowing that and knowing how. So neither the forces of production nor the relations of production nor any social existence that includes the two can fail to involve quite a number of forms of consciousness, and none of these can cause whatever forms of consciousness are involved in it. However, this difficulty disappears as soon as we explicitly introduce the time dimension. Social existence, in all its aspects, at one time t_1 is a distinct occurrence from all forms of consciousness at any later time t_2, and may well be causally related to it. If the forms of consciousness are the same at t_2 as at t_1, we can say that social existence is a sustaining cause of them: if they change, then it can be a cause of those changes.

Though this resolves the problem of getting logically distinct causes and effects, there may be another objection. In so far as consciousness is a component of social existence at t_1, consciousness at t_2 will be just a later phase of the same thing or activity, and we do not commonly say that a persisting thing *causes* its own later existence, or that an earlier phase of a continuous process *causes* a later phase of that same process.

Hence it might be argued that if C_2 is a later phase or continuation of C_1, we should not say that a whole consisting of C_1 and other things causes C_2. Admittedly we do not speak in the ways mentioned. But I think we might well extend the concept of causing to include these, since in all important respects the relations between earlier and later phases of a thing or process are just like the relations that we unhesitatingly accept as causal. But in any case nothing turns upon the use or non-use of the word 'cause'. Marx's word, after all, was 'determine', and it will be no great loss if we say that this determining is not exactly causal, but is just like causal determining in all important respects.

We encountered another form of the logical connection problem with regard to explanations of the second group. If we explain the actions of people of a certain class by reference to a class ideology in which they share, this is not a causal explanation of what is commonly taken to be the typical Humean form. It is not a purely contingent regularity of succession, discovered by observation of a number of particular sequences that instantiate it, that people's having a certain ideology leads to their acting in such and such a way. Rather their having this ideology makes their acting in this way (relatively) rational, and rationally intelligible; we should expect that if they have this ideology they will act accordingly; and we should expect this independently of any particular observations. Typically, this will be because the ideology shows the actions in question in some kind of favourable light, as being natural, or part of the divine plan and order, or as being necessary for some end which is itself also shown as desirable, or perhaps as historically required and inevitable. The ideology may in some such way as one of these itself motivate actions, or it may merely remove inhibitions that the agents might otherwise feel, as the ideology of *laissez-faire* removes inhibitions about cut-throat competition and the driving of hard bargains, and a revolutionary ideology may remove inhibitions about killing and destruction.

There are several points to be made here. First, the fact

that ideology-action is not a non-expected regularity of succession is irrelevant. It might be expectable and expected because it exemplified a well-known general pattern of human behaviour, but it could for all that be a purely contingent regularity satisfying the most extreme Humean demands. Secondly, we seem to have here a variant of the standard 'logical connection argument' which purports to show that intentions cannot be causes. The simple version of this is that A's intending to do B can be adequately described only as his intending to do B, that is, in a way that logically involves the notion of A's doing B. So A's intending to do B and A's doing B are logically connected, so that the one cannot be even a partial cause of the other. But a logical connection of this sort will not prevent A's intending to do B and A's doing B from being distinct existences in Hume's sense—they can even be separated in time—so that this is no obstacle to their being cause and effect one of another. But some who have recognized this (notably von Wright and Stoutland) suggest that there is another sort of logical connection which makes a purposive explanation of action non-causal. Thus von Wright sees a purposive explanation as simply turning upside down a practical inference which when fully formulated is logically binding. Essentially, a person who fully intends to do something and believes that the time for doing it has come logically cannot fail to set himself to do it: if he fails to do so, it will show that he either does not fully intend this or does not believe that the time has come.

Any force that this argument has lies in its claim that A's fully intending to do B now and his now setting himself to do B are not distinct occurrences, that they are just the same thing described in slightly different ways. But if so, not only can the one not be the cause of the other, it cannot explain the occurrence of the other in any way at all. If we go beyond the bare report 'A fully intends to do B now' and put in A's further reasons we might be said to explain his action in the sense of saying more fully what action it was: it was doing B as this or for the sake of that. If we include these features in the action, then indeed *A*'s having these reasons is not a distinct existence from and therefore not a cause of the action, and equally there is no explanation of the occurrence of the action, though the action itself is explained or explicated in the sense that it is unfolded, set out more fully. On the other hand, it seems impossible to deny that there is such a thing as a pre-formed intention, and hence a state of an agent's having such an intention before he performs the related action which, being temporally separate from the action, must be a distinct occurrence. If so, it is hard to deny that a similar state might occur contemporaneously with the action, and yet be still an occurrence distinct from that action, or, if the action itself is so defined as to include this state, distinct from certain parts of the action. And then not only could the having of a pre-formed intention be a (partial) cause of the whole action, but also the contemporaneous having of the intention could be a (partial) controlling cause of those other parts of the action. In this way the having of intentions could, for all that the logical connection argument can show, be causes of actions or of parts of actions, and so could figure in causal explanations of their occurrence, even if there is *also* the kind of explanation for which von Wright argues which explains, non-causally, not the occurrence of the action but the action itself just by describing it more fully. But what we must not do is confuse these two, and suppose that we have found a non-causal kind of explanation of the occurrence of an action.

This conclusion is not undermined by the consideration that an existentially separate—for example pre-formed—intention is also rationally connected with the action, that the having of it makes it so far reasonable for the agent to do what he does. It is a contingent fact that people do act on the whole fairly reasonably, even though this fact has been taken up into our concepts of wanting, deliberating, choosing and so on. Nor can we build anything on the admitted awkwardness of such a remark as 'He was caused to do *B* by his having earlier formed an intention to do *B*'—a formula which is inappropriate just because it suggests a cause bearing upon the agent from outside. Nor can it be argued that the relation in question cannot be causal on the grounds that there seems to be no way of reducing it to a mechanical process, or that there is no good evidence that what is going on here is deterministic.

All this can be transferred to the realm of ideological explanation; the having of an ideology, like the having of an intention, combines elements of desire or valuing and belief, and it similarly makes a corresponding action so far reasonable. Also, the ideology, like the intention, can figure in two different sorts of explanation. It can be taken in something like von Wright's way as not existentially distinct from the action, but as helping to make it the sort of action that it is. For example, the passing of a bill, and so the making of a law, by a parliament in what Marxists would call a bourgeois democracy is an occurrence for the adequate description of which we need to mention the roles that the various agents see themselves as playing, the purposes of which they are conscious and the powers they take themselves to have. But the ideology can also figure as a partial cause in a causal explanation. If we ask why it came about that there is now this new law, it may well be correct to refer to the ideology shared by most members of the parliament and by many of their constituents as a partial cause of this event.

We can now see that the dilemma that I posed earlier, that an explanation from inside an ideology would be biased and distorted, and that one from outside would ignore something essential to the social phenomena, is almost wholly spurious: neither horn is either hard or sharp. An adequate description of a social phenomenon must indeed include an account of what the agents take themselves to be doing. In one sense they cannot be wrong about this. But it is also possible that their behaviour has some partial causes of which they are unaware, that it will have effects that are not included in their purposes, and indeed that it may have an unknown function in the sense that these unknown partial causes include the fact that such actions tend to produce these effects which were no part of the agents' purposes. Thus there is a sense in which the agents can be wrong about what they are doing. But a full description of what is going on must and can take account of both aspects, both of how their actions appear to the agents and of what is not apparent to them: it can recognize distortions as part of what is there without itself becoming distorted. Accounts of a social phenomenon from inside and from outside the relevant ideology, far from being both inadmissible for different reasons, are both admissible and both contribute to understanding, though in different ways.

This internal/external contrast does not quite coincide with the non-causal/causal one that I used above. It is true that the external account will be in several important and possibly complex respects a causal one. I have suggested that the action (A) will have some unpurposed effects (E) and some unknown-to-the-agents causes (C), and further that C may include the fact that actions such as A have regularly had effects such as E: such actions may have been fostered by their tendency to produce certain results: in this sense they may have a function unknown to the agents. It is equally true that an internal account may be non-causal, it may simply describe actions in which beliefs, purposes, etc., are constituents and so display whatever rationality they have. But there is no reason why an internal account should not also be a causal one, why it should not treat beliefs and purposes, and also quite different sorts of things, as causes of actions or of other effects that are existentially distinct from those causal factors. I mentioned at the beginning of this paper an explanation of inflation given by bourgeois economists: this would clearly be a causal explanation. Being internal to an ideology, such a causal explanation will no doubt be inadequate in some ways, it will leave out relationships which from outside the ideology can be seen to be important. But it is not necessary that the causal statements made within an ideology should be false: many or all of them may correctly or fairly correctly identify some partial causes, subject only to the limitation that the items identified as causes may themselves be seen, from within the ideology, in a somewhat distorted way.

I come back now to my first group of questions, whether all thinking about society is ideological, whether there is therefore a general sociology of knowledge, and whether ideologies have to be false or distorted—or, if they are made so by definition, whether this feature necessarily accompanies the other features that constitute an ideology.

Now it is obvious that for an ideology to perform the sort of function that it is supposed to perform, it must include value-components. Also, it is very natural that what we may be able in principle to distinguish as factual and evaluative components should be almost inextricably combined in the content of the ideology. To put it very crudely, a system of concepts, beliefs and values will be more stable and more effective in controlling and justifying conduct if the evaluative aspect is wrapped up in the concepts and beliefs, if those who have the ideology see things-as-they-are as exerting certain pressures on them. Slanting the news reports is a more potent form of propaganda than printing a rousing editorial alongside a neutral and objective report. For similar reasons, we are less likely to find value merely added to descriptive elements like a surface colouring, than incorporated in descriptions by the addition of fictitious elements and the omission of much that is real, the magnifying of some real elements and minimizing of others, and so on. Ideologies being what else they are, it is to be expected that even in their descriptive aspect they should contain a good deal of error and distortion; and yet it is also necessary for their operation that they should have verisimilitude, so that they cannot afford to be completely false.

I said at the start that I was not primarily concerned with the question how widely applicable the notion of ideology is. This is an empirical question, and can be properly settled only by detailed studies. Still, it may be legitimate to remark that on general grounds one would expect thinking about society to be largely ideological. The basic everyday concepts are those used by men who are practically involved in social action, and who will naturally have a need to justify their actions, to explain them as rational to themselves and others, to see them as having places in a value-pervaded scheme of things. And even if social theorists are not themselves, even in a minor way, men of action, they will unavoidably start from and be influenced by the thinking of those who are. No doubt one should take account account here of the Marxist view that the thinking of those who are on the side of ultimately inevitable progress, whether we call it an ideology or not, is likely to be truer and less distorted than that of those whose social role is conservative or reactionary. But all I can say here is that I see neither any *a priori* plausibility in this thesis nor any *a posteriori* confirmation of it.

But although we have some grounds for expecting all thinking about society to be to some degree ideological, we have none for embracing a general doctrine of the sociology of knowledge, for saying that all truth about society is socially relative, that no proposition about society is simply true, but that even the most favoured items are true only from this or that point of view, or only within this or that ideology. It is easy to show that this extreme doctrine is self-frustrating, that taken seriously it would prevent even itself from being enunciated. But equally there is no need to adopt it, since the general grounds that we had for thinking that all ideologies involve some distortion are also grounds for expecting them to contain a fair amount of simple truth as well.

My main message, therefore, is one of tolerance. I have found no philosophical reason for scepticism either about the very concept of ideology or about any of the several kinds of explanation into which it may enter. In conclusion, I want to sum up these possibilities in broad outline and see how they fit together.

An ideology (I) makes reasonable certain kinds of action (A). Consequently, the possession of that ideology by the members of the society in question is a partial, favouring, cause of actions of these kinds. But also actions of these kinds (A) are, in the actual environment (E), helpful for the continuance and flourishing (F) of a society of that sort. And these two facts together explain causally, in the usual evolutionary way, the growth and persistence of that ideology. I's being a favouring cause of A, together with A's being in E a favouring cause of F, together constitute a favouring cause of I itself. (The apparent circularity in this outline account disappears as soon as we introduce the time dimension.) We find, then, that though not everything that can be called ideological explanation is causal, it is within a complex of causal relationships that we can understand both the occurrence of ideologies and their various contributions to explanation. Also, we might develop the pattern just sketched by bringing what Marx called forces of production in under E and what he called relations of production as an extension of A. We can then see that *wherever this pattern is applicable*, it will be plausible to regard the forces of production in E as the independent variable, and the relations of production in A as the factor more immediately controlled by it, I being more remotely controlled. This yields another sense in which, without making ideology epiphenomenal, we can still say that it is social existence that determines consciousness, rather than the reverse.

Comment: Ideology and the Modes of Explanation

BY RENFORD BAMBROUGH

The many merits of Mackie's paper are intrinsic and evident. Its major deficiency is relative to the needs and duties of a commentator. His paper contains neither any mistakes that it is the duty of a commentator to correct nor any exaggeration or rashness of generalization that it is the duty of a commentator to moderate. He is neither wild nor woolly. He has offered what seems to me to be a correct and useful characterization of some things that might be meant by the term 'ideological explanation', and has convincingly resolved what he calls the spurious dilemma according to which every account of individual or social action must be either too external and detached to achieve understanding of agents and societies, or too internal and partisan to be objective and unbiased. On his chosen ground—and it is extensive ground—Mackie has done so much so well that there is little for a commentator to do but to endorse his findings and then, after the manner of commentators in such cases, to talk about something else.

Before I leave Mackie's paper, however, I should like to raise one or two cavils and queries which will pave the way to the different but related matters that I want to put to the question.

1. On p. 186 Mackie asks whether all thinking about society is ideological. Since he has already adopted a definition of 'ideology' according to which every ideology is false, he could take a shorter way than he does with this question. It could not be known that some or any or all views of society were ideological in this sense without its being known that the views in question were false, and the knowledge that a given view of a thing is false is knowledge of some truth about that thing. So that we shall at any rate have no application for this definition of ideology unless we can know some truth about society, and as soon as we know some truth about society we know that not all thinking about society is false and hence that not all thinking about society is ideological. I have heard of a medieval historian who believes that the Middle Ages are unknowable, and whose method of supporting this view is to demonstrate the falsehood of every thesis about the Middle Ages that any other scholar may put forward. But it occurs to me that this may be the sort of incoherence that Mackie is diagnosing when he rejects 'a general doctrine of the sociology of knowledge' as 'self-frustrating' (p. 196), and if so then my first query or cavil lapses into another endorsement.

2. Though Mackie defines an ideology as involving falsehood, he suggests on pp. 195–6 that an ideology 'cannot afford to be completely false' since it is necessary for its operation that it 'should have verisimilitude'. Earlier (p. 194) he has allowed that an ideology may include true causal statements. Here again I accept Mackie's conclusion but I suggest that there is a deeper reason than he gives for it. Every ideology will contain some truth, not or not only because of the prudence or cunning of the ideologist, but rather and mainly by a logical necessity. For as Mackie himself makes clear, no 'system of concepts, beliefs and values' can qualify as an ideology unless it has wide scope and operates on a large scale. Even if those words from the first line of Mackie's paper had not made this point, it would be evident from some words that he uses later in the same sentence: it is in terms of such a system, he says, that a group of believers in an ideology 'see and understand their own position in and relation to their social environment and the world as a whole'. Now nothing could be recognizable as an account of anyone's or anything's 'position in and relation to the world as a whole' and still be 'completely false'. And this is shown by a consideration that has nothing specifically to do with ideology. No account of anything that is large and complex, however much error and falsehood it may contain, can be 'completely false', since in order to be seen to be representing or misrepresenting its object it must be seen to be referring to that object, and it could not be seen to refer to an object concerning which it contained no truth at all. If my biography of Napoleon contains too few truths about Napoleon it is not a biography of Napoleon. If my map of Treasure Island is literally *all* wrong it is not a map of Treasure Island. If nothing that I say about Mackie's paper is true, then I have not offered a commentary on Mackie's paper.

This takes us to a further point on which Mackie offers a correct conclusion but supports it by only some, and not the strongest, of the reasons in its favour. That everyone concerned to do for himself or his group what, according to Mackie and Marx, an ideology can do, will be liable to error is clear and obvious; and no doubt many of the typical errors of ideologists are attributable directly to the bias on which Mackie puts the stress. But several other sources of error need to be mentioned, none of which has anything specifically to do with ideology. Mackie would not of course deny, though it might have been better if he had asserted, that some of the errors made by ideologists would simply be errors, attributable to the general human fallibility of those who make them: misperceptions, miscalculations, confusions such as might occur in any sphere. Others will arise from the scale of operations to which Mackie and I have drawn attention, and from the complexity of the situations, facts and circumstances that social inquiry has to deal with. Where an inquiry has to cope with such scale and complexity, whether or not it involves the further problems associated with man, mind and society, or with judgements of value, distortion and misdescription and misrepresentation are inevitable.

The position we have now reached is this: an ideology, like any other large-scale 'system of concepts and beliefs', whether or not it is a 'system of concepts, beliefs and values', will necessarily contain a good deal of truth, and will inevitably, though not by a logical necessity, contain a good deal of falsehood and distortion. All this applies to large-scale inquiries of any sort, whether they are theoretical or practical, and whether or not they involve mind and value. These facts are implicitly recognized by many of our social institutions and private practices. The adversary system in courts of law, the two-party system of government, the practice of inviting commentators to reply to papers at philosophical conferences, and audiences to comment on the commentaries—all these are devices for ensuring as far as possible that we can combine an aspiration after scope and comprehensiveness of inquiry and understanding with the minimum of misrepresentation and distortion. It is on the same foundation that Mill builds the case for liberty of thought and speech and action: that error may be exposed, and truth revealed. These steps are designed to ensure that the partiality that is *parti pris* will not necessarily enslave us to the partiality that is incompleteness of coverage or grasp or attention.

3. My last direct comment on the detail of Mackie's paper concerns a related point. On p. 192 he writes: 'It is a contingent fact that people do act on the whole fairly reasonably, even though this fact has been taken up into our concepts of wanting, deliberating, choosing and so on.' It is not clear to me what is or could be envisaged as an alternative to the supposed contingent state of affairs here described, or to the concepts whose character is in question. People who did not act on the whole fairly reasonably would not be recognizable as people. Nor is it clear how one could describe alternative beings such that a recognition that they did not on the whole act fairly reasonably could be combined with a recognition that they did *act*, and in ways and senses that made their conduct a proper object of assessment by standards of rationality at all. It is a contingent fact that there are *animalia rationalia*, and a contingent fact that there are any human beings; but it does not seem to be a contingent fact that a description cannot be a description of human beings unless it is also a description of *animalia rationalia*.

For the rest, what I should like to offer is a supplement to rather than a criticism of Mackie's paper, and it happens that it relates to a theme that has recurred in the papers of this conference and in our discussions of them. Mackie implicitly allows on pp. 191–2 of his paper, in his references to von Wright, that something may qualify as an explanation in a good sense of the term even if it is not a causal explanation, but is rather an unfolding of the internal structural complexity of, for example, an intention or an institution or an institutional procedure such as the passing of a parliamentary bill. I hope I may be forgiven for suspecting, on the evidence of tone rather than of specifiable content, that Mackie thinks of such non-causal explanations as belonging only to the antechambers of the palace where causal explanation sits enthroned. Certainly in other papers and discussions, both at this conference and in other places and times, the impression has often been given that explanation must be regarded as a unitary or homogeneous concept, and that the outlying instances or types of explanation must either be construed as conforming to the causal pattern or set aside as being explanations in some other sense of the word than the one that philosophers of science have to do with. An eminent logician has been quoted as saying that proof theory does not have to give an account of the proof of a pudding or of proof spirit.

There is a danger here of falling into merely verbal wrangles, but there are also serious issues that must be faced. One of them can be at least raised by asking whether the phrase 'unitary sense' itself has a unitary sense, and if so, in what sense of 'unitary sense' it has one. From the time of Aristotle it has been known by some philosophers and forgotten by others that there are many ways in which a term that is not straightforwardly univocal may nevertheless avoid being merely equivocal. Among the familiar classical paradigms are 'medical', 'healthy', 'being' and 'good'. There is still much controversy about these matters, but it is at least very widely recognized that the *epistemic* terms—*knowledge, truth, reason, proof* and their kindred—are strong candidates for an analogous treatment. Now *explanation* is clearly a scion of this stock, and one might therefore expect it to show the versatility and internal complexity that are characteristic of the whole family. To explain is to increase *understanding*, to remove *perplexity* or *puzzlement*, to diminish or banish *surprise*. As Aristotle said, the geometer, who understands the matter, would be amazed if the diagonal of the square were *not* incommensurable with the side. Now to explain to somebody who is surprised at this why he should not be surprised is not to give him a causal explanation but it is still to give him an explanation. Philosophers should be readier than they often are to recognize that the plurality of modes of understanding is matched by a plurality of modes of explanation. Besides explaining why a bomb explodes and a car does not, or (on a particular occasion) why a bomb does not explode and a car does, we can also explain motives, policies, riddles, allusions, mistakes, people, maps, paintings, plays, poems, proofs, arguments, and explanations. Many of the items in this list, which is selective almost to the point of being casual, are capable in different or even in similar contexts of admitting explanations of two or more of the many types of explanation that need to be distinguished if there is to be an adequate description of 'the concept' of explanation.

If we bear all this in mind when we ask Mackie's question (p. 186) 'What kinds of explanation are possible in the social field?' we shall save ourselves from some oversimplifications and distortions to which answers to such questions are easily liable. While it is a commonplace that the social sciences are not to be forced into moulds made for the physical sciences, the commonplace is often a parrot cry, not backed by anything like a full understanding of the truth that it expresses. It is often accompanied by the idea that there is an alternative but still fairly simple and unified model for the social sciences and the explanations they aspire to give. Even my short list of items that may be proper objects of explanations is long enough to show that explanations of many kinds may be expected to occur in social inquiry.

I speak of social inquiry in order to make the field wide enough to admit what is often excluded by those who are bent on securing autonomy and specialist professional status for sociology or social anthropology or social psychology, or for a conjunction of two or three of them, under the banner of social science. Even in the short time that remains to me I can speak of two such cases of banishment, and all the more conveniently because the crimes with which the exiles are unjustly charged would find no place in a calendar rewritten on the lines suggested by my discussion.

The first is history. It would be remarkable, if it were not so typical of the departmentalism of philosophical studies, that in a conference on the concept or concepts of explanation there has so far been no reference in any paper or discussion to the idea of historical explanation. When historical explanation *is* referred to by philosophers they tend to assume that all historical explanations are causal, too. But another reference to my list on p. 203 will show that many other modes of explanation are employed and need to be employed in the writing of history. Even in some places where causal language is naturally and properly used, the type of explanation that is being offered is not causal in the sense favoured by most philosophers of science. To seek the causes of the French Revolution or the First World War or the Renaissance or the Reformation is to explore connections and cross-references in complex patterns of particularity in a manner analogous to that involved in the critical analysis of a novel or a symphony or a classic work of architecture or painting or philosophy. In all such cases an explanation may consist in a re-presentation of what is already before us rather than in a reference to something external to it. And in the historical instances the explanatory technique may be just that technique of orderly narrative which is most liable to be contrasted with explanation when philosophers are allowed to interfere with history or historians.

My second example may be thought to be more dangerous still. It is literature, and by literature I mean what one eminent sociologist is in the habit of calling 'belletristic literature', in case it may become confused with 'the literature'—a very different kettle of red herrings.

The understanding of man and society has been advanced by works of fiction as much as by works of fact, and our discussion today can help us to see why it should be so. In *War and Peace* or *À la Recherche du Temps Perdu,* or the works of Shakespeare or Dickens or Swift, we find presentations of 'concepts, beliefs and values', and in terms of such works men can 'see and understand their own position in and relation to their social environment and the world as a whole'. That these are works of fiction means that they contain some falsehoods. That they are recognizable as pictures of man and the world and society means that they also contain much truth. What is more, there are two ways in which the falsehood that they contain contributes to the apprehension of the truth.

The first way is suggested by the parallel with ideology that I have marked by quoting those phrases from Mackie's paper. Proust and Dickens, like Marx and Machiavelli, (and Freud and Plato and Spinoza), and like all the other thinkers who have most often been accused, relevantly or irrelevantly, of being 'unscientific', bring features and aspects of our world into relief by selecting them for emphasis at the expense of other elements in that comprehensive truth after which no one work or author could aspire without lapsing into banality or causing bewilderment.

The second way is classically followed by Swift and Orwell, though it should not be thought that it leads only to satirical objectives. It consists in devising imaginary 'objects of comparison' or objects of contrast to set beside the familiar realities that we are most likely to misunderstand because they are most real and most familiar. When Gulliver has told us of the pygmies and giants he saw on his travels we have seen himself and ourselves as giants and as pygmies and are ready to see ourselves as men, but with two new perspectives on our place in the scheme of things actual and possible. When we come back to 1973 from *1984* or back to England from *Animal Farm* the here and now looks different while still looking the same, as England looks most foreign and most English to an Englishman driving back from France or landing from New York.

Comment: Ideology and Metaphysics

BY MARTIN HOLLIS

The notion of Ideology is a worthless tool, if allowed to explain too much, and Mackie's attempt to make it limited but precise will have stirred sympathy. Paradoxically, however, his way of doing it seems to share the fate of King Midas, by turning *all* explanation into ideological explanation. Having tried to show why, I shall make some fleeting remarks about the comparison of categorical framework and then suggest that ideological explanation be confined to the social causes and functions of intellectual error. If my canvas is larger than Mackie's, it is partly because his brushwork defies complaint. But it is also my contention that his unstated assumptions beg some crucial questions and I shall end with a plea for intolerance.

Mackie's initial definition of Ideology is in effect a compressed history of that difficult notion. There seem to be three strata to it. The first, from pre-Marxian socialism, involves an analysis of society in terms of economic class, a social structure determining an intellectual, legal, and moral superstructure, and the idea that ideology distorts. The second, more typically Marxist, puts false consciousness to work mystifying and so perpetuating a social system and thus undermines the distinction of structure from superstructure. The third, abruptly modern, extends the idea to other social groups, perhaps even a whole society, and throws in any latent functions which ideologically prompted action may have. The pre-Marxian stratum thus contains a fairly precise political theory to end political theories and so gives a fairly precise sense to 'ideological explanation'. But its shortcomings are plain and there is no resisting the broader and more powerful thesis of the Marxist stratum. Yet, with consciousness an ingredient in the identity of social groupings and social theorists interested in more than capitalism, it is equally hard to resist the third stratum. By now, however, 'ideological explanation' has lost its precise sense. In upshot the definition of Ideology fits any religious or metaphysical account of man in his enviroment and so, apparently, any scientific or even perhaps common-sensical scheme of things. If so, Mackie's attempt at a non-proliferation treaty cannot succeed.

Mackie offers the treaty by making assumptions which

general theories of conceptual imprisonment usually deny. He speaks throughout as if he had a neutral standpoint for judging the facts of the world and neutral criteria for assessing the merits of ideological explanations. He holds to a distinction between the intellectual and the social (despite an intrusion of consciousness into social existence early in the paper), without which his conclusion lapses. He takes it to be easily shown that an extreme doctrine of the sociology of knowledge is self-frustrating. Given these assumptions, the notion of Ideology has to wear a collar and tie and behave in a decently British manner. But they are, I submit, gracefully disingenuous. To inject a teasing note into our stern discussions, if British common sense is not an ideology, the paper shows broadly that foreigners are funny; if it is, the paper is a walking illustration of the conceptual imprisonment it assumes away.

To introduce the point of the tease, we may ask whether these are ideological explanations:

1 An explanation of neurosis by Freudians, using concepts from Freudian psychology.
2a An explanation of Chinese interest in acupuncture by reference to Chinese medical tradition and current lack of resources to spend on drugs and machines.
2b An explanation by an acupuncturist of Western dismissals of acupuncture in terms of Western beliefs about bio-chemistry.

Mackie calls it an empirical question, but detailed studies will not tell us until we know what we are looking for. He will perhaps retort that we do know—we are looking for partly false beliefs with latent social functions; what is not clear is simply whether Freud is metaphysics and acupuncture magic. But, waiving the point that all action-guiding beliefs have latent social functions, I would think this unduly curt without something to show that one man's metaphysics is not another man's science and one man's magic not another man's technology. Otherwise we shall have to conclude that all explanations are ideological, in the absence of an uncontentious criterion of objectivity. Since schemes of belief include their own differing accounts of objectivity, Mackie's implicit neutral observer, who so deftly unhorns our dilemmas, does so by begging our questions.

To emphasize that the issue is the structure and not just the scope of ideological explanation, let us focus on the clause in the definition which reads 'at least some of the beliefs and concepts in the system are false, distorted or slanted'. Why precisely might a bourgeois economist's view of inflation call for ideological explanation? Suppose he typically blames inflation on cost-push driven by demands for higher wages. The truth or falsity of this particular reckoning is not crucial, since an ideology can include much truth and its antidote much falsity. Presumably the question turns on further beliefs, which the economist would cite as his reasons. These further beliefs either give good reasons or they do not. If they do not, we have a prima facie case for a different type of explanation of why he thinks they do. If they do give good reasons, then the question of their truth arises in its turn. The process continues until we reach fundamental beliefs which are false or held for bad reasons. We now have a prima facie case again. It is only a prima facie case but it raises several points. In making it, we have credited ourselves with objective criteria of truth and of rational belief (to say nothing of the power to apply them correctly) and we have assumed that true beliefs, rationally held, are their own explanation. Had we taken different criteria or not made that assumption, we would have a different account of ideology. Whence comes our right to be sure?

The problem posed is familiar, so I shall risk putting it in a less familiar way. It arises in social anthropology in comparing categories of thought among cultures, where, by historical accident, it often takes a revealing form. The theorist has often been reared on Positivist criteria of truth and rational belief which lead him to draw that sort of sharp distinction between science and other modes of understanding. He then takes other cultures to be pre-scientific and explains their distorted view of the world partly (1) by ascribing to them a prelogical mentality or other idiosyncratic schemes of symbolism, partly (2a) by their social systems and partly (2b) by exhibiting the latent functions of the distortion. It then strikes him that Positivism is itself unscientific by its own criteria. Hence he concludes that his own culture is to be explained ideologically too and that Objectivity is a myth or rather a set of overlapping myths. Having thus piled Pelion on Ossa, he can survey the philosophers on Olympus with amusement, at any rate provided he conveniently forgets to apply his conclusions to himself. When asked what he has discovered, he is prone to reply, 'that all truth is relative'. But, spotting the need for an exception in his own case, he soon amends his discovery to read 'all truth is relational', to allow second order judgement of first order relativity. He now cannot explain, however, why the second order is less relative than the first. Nor does it help him much at present to call in the philosophers. He has gone too far to accept 'British common sense' as a neutral standpoint and any forms of metaphysics or philosophy of science which are finally descriptive of current categories merely pose his problem more philosophically. His plight seems hopeless, when even such a champion of the grand tradition in philosophy as Stephan Körner declares even the law of non-contradiction parochial in principle, remarking:

> . . . the statement that *the* proposition *not* (g_o *and not-g$_o$*) is logically valid in *L* and *I* is either false or an elliptical way of stating that of the *two* different propositions represented by '*not* (g_o *and not-g$_o$*)' in *L* and *I* respectively, the first is valid in *L* and the second is valid in *I*. If the view that the *one* proposition *not* (g_o *and not-g$_o$*) is valid in *L* and *I*, and thus incorrigible with respect to any categorial framework with underlying *L* and *I*, breaks down because there is no one such proposition, it follows that there is no one logical proposition *not* (g_o *and not-g$_o$*) which is incorrigible with respect to every categorial framework. . . .[1]

[1] S. Körner, *Categorial Frameworks*, Blackwell, 1970, p. 18.

With the suggestion that truth is itself relative to a framework, we are bound to assert that all explanation is ideological; and, in so doing, to reduce all inquiry to absurdity. This result cannot be allowed to stand.

The problem in more familiar form is implicitly posed by Mackie's initial definition, which is so broad that an ideology differs from a rational or scientific scheme only in that it involves greater distortion. The distortion, as Mackie characterizes it, is not peculiar to any special kind of thinking but relies solely on a distinction between conceptual imprisonment and an objective standpoint. But any general theory of objectivity also falls within the definition of ideology. One such theory is presupposed by 'British Common sense'. It is not the only theory but that fact merely makes visible what would still be true even if it were, namely that all explanation involves presuppositions which are ideological by the definition given. Consequently the extreme doctrine of the sociology of knowledge needs to be taken seriously, not so much because it is tenable as because the exact line of defence against it affects all the familiar demarcation disputes which absorb philosophers of science. Meanwhile, I agree with Mackie, we must reject the threatened conclusion that all explanation is ideological. That conclusion has to be shown to involve two mistakes, one in equating the possession of an ideology with all other forms of conceptual imprisonment, the other in holding that no truths are incorrigible with respect to all categorial frameworks. Otherwise the day of the sceptic will have finally come.

I have no space here to discuss the limits of conceptual imprisonment. I just want to submit for discussion the thought that, fortunately, Körner cannot mean quite what he says in the passage just quoted, since he could not have uttered the passage at all, if what it seems to say is true. His words evidently belong to the language once called in hope *characteristica universalis*, a language which allows the comparison of lesser logics just as it allows a discussion of various geometries. For, if the passage is not in *characteristica universalis*, it is unintelligible and, if it is, it is false. In brief, I unite with Mackie and Körner in denying that all truth is relative. But, whereas Mackie seems to hold that some facts are given independently of all frameworks, I unite with Körner in thinking truth relational. Yet, unlike Körner, I submit that, if relational is not to collapse into relative, there must be categorial propositions incorrigible with respect to all frameworks. At any rate, I shall assume it, in order to work back to a definition of Ideology.

With further apology, I must also forgo the next step in a full argument, which would be to persuade the social anthropologist that science, religion, and metaphysics are crucially similar intellectual schemes for making the world intelligible. All try to understand a world underdetermined by experience—a platitude rendered harmless by blocking the extreme sociology of knowledge which earlier seemed to follow from it. To make amends to Körner, I would now lend the anthropologist a copy of *Categorial Frameworks*, which sets out the kind of understanding most suggestively. For the present, however, we must take it as read.

We left the bourgeois economist holding fundamental beliefs which were either false or professed for a bad reasons, in order to create a prima facie case for a different type of explanation. Perhaps we can now suggest where the difference in type of explanation lies. Categorial frameworks are *intellectual* structures and, while the economist is operating successfully within his framework, there is nothing for ideological explanation to bite on. For, in as much as his beliefs are true and his reasons good, he is indeed understanding the world. Moreover, where he does have a bad reason for a belief, the ideologist's interest need not focus there; if he also has a higher order canon of inference or evidence or the weight to be attached to evidence, which licenses appeal to the bad reason, then the ideologist must fasten on the canon. Ideological explanation is a diagnosis of error and, were we to admit the possibility of a true ideology, it would cease to have any value.

Positively, the diagnosis must be a social one. The economist must be in error *because* he is a *bourgeois*. Both emphasized words give trouble. I agree with Mackie that 'because' must be causal or functional otherwise there will have been no reference outside the system of beliefs. But I am not convinced that Hume is in any position to say so. Admittedly the introduction of a time dimension is a neat touch, since any series of events can always be continued in more than one way under some description, thus ensuring that the connection of cause and effect is contingent under that description. But it does not follow that, in general, a state's being what it is at t_1 is logically distinct from its being what it is at t_2—even for Hume predicates which presuppose regularity apply only where there is general regularity, even if particular exceptions are allowed. I take it, however, that Mackie is trying to show Hume not so much right as not wrong for a fashionable reason and I have no quarrel with that. Meanwhile the causal 'because' must be construed both ways—in pre-Marxian terms the economist would not be in error, were he not bourgeois, and in Marxist terms he would not be bourgeois, were he not in error.

What saves the Marxist version from circularity is its insistence on classes defined in relation to the means of production. With so firm a social map it is safe to add that shared consciousness makes the difference between members of a class and potatoes in a sack. But a more plural social analysis can lead to circularity. If workers are merely a social group, then so are liberals, dentists, and pigeon-fanciers. Each of these groups has an 'ideology', as Mackie originally defined the term, and that 'ideology' is, duly, a system of concepts, beliefs, and values, which unify each group in virtue of shared perception of a common interest. An 'ideology' becomes the theoretical unity of a social group and a social group becomes whatever has an ideology. In this already circular account the word 'social' is doing no work and an 'ideological' explanation becomes merely any account of how men with a perceived common interest act and justify acting in pursuit of that interest. Since 'interest' here can include anything from robbing a bank or cleaning up television to worshipping God or playing Go-moku, we are back where we started with ideological explanations which explain too much. The bourgeois economist must belong to a social group which is primary in a true social theory, shares an ideology because it is there and continues to be there because it shares an ideology. In sum, the distinction between intellectual and social systems is crucial and the definition of 'ideology' must advance a political thesis about the existence and function of intellectual error.

An ideology, then, is a system of concepts, beliefs and values which is characteristic of some primary social group and in terms of which the members of the group claim to see and in fact misunderstand their own unity and the world as a whole, and explain, evaluate, and justify what they do because they belong to the group. Thinking in terms of this system unites and weakens the members of the group by helping to perpetuate a structure contrary to their interests. Such a definition widens a classically Marxist one by allowing that economic classes need not be the, or the only primary, groups. But it makes no other concession. In particular 'ideological explanation' ceases to refer to (1) an explanation given within and in terms of an ideology; and becomes strictly an account of relations between the social and the intellectual, showing (2a) why members of a primary group hold the beliefs they do or (2b) why given beliefs have the manifest and latent social functions they do in the lives of members of a primary group. I take it that neither Freudians nor Chinese nor acupuncturists nor Westerners nor biochemists form a primary social group.

As a final footnote to this account, an odd result is that only mystified workers (or other exploited primary groups) have an ideology. Mystified members of the ruling class have a set of beliefs which advances their interests and unmystified members of all groups do not have an ideology at all. I take this as a sign that the idea of 'interest' is more important and trickier than has emerged, implying that the account is still vague and incomplete. Meanwhile, however, I am consoled to note that, on Mackie's definition, all these groups have an ideology *except* mystified members of the working class, whose beliefs do not advance their interests. But, as we already have enough to discuss without this, my advice would be 'Praise the Lord and pass the opium!'

Reply to Comments

BY J. L. MACKIE

Since so much of Mr. Bambrough's paper is either complimentary or complementary to my own, I am content to leave the few substantial disagreements between us to the arbitration of our readers. But though it may seem churlish or dull to make a solemn reply to Dr. Hollis's light-hearted raillery, I must take up two groups of his comments which, by combining rhetoric with imprecision, are liable to confuse and mislead.

First, there is the complaint that my account of ideology is all-embracing. But even if all explanation (or perhaps all explanation in the social field) were ideological, this would not make the notion of ideology or of ideological explanation worthless. There might well be features which were common to all (social) explanations, to which it was none the less worth while to draw attention. In any case, I have distinguished three different sorts of explanation each of which might be called ideological, and there would still be a need for these three separate notions even if an umbrella term covering all three were of little use.

Consequently Hollis's proposal to define 'ideology' more narrowly is poorly motivated. He suggests two restrictions: the first, that only a social group which is 'primary in a true social theory' should be allowed to have an ideology, the second, that an ideology must *weaken* the members of the group whose ideology it is by helping to perpetuate a structure *contrary* to their interests. Of these, the first may be useful, in that the ideologies of primary social groups may be of such distinct importance as to deserve special consideration and hence a special name. But the second (which would forbid Marxists, for example, to speak of a bourgeois ideology) would deprive us of a useful general term. If we accepted it, we should at once need to introduce new terms (*a*) for the very class of things that Hollis's restriction excludes, and (*b*) for the wider class that includes both these and what he would call ideologies. More disturbing is his complaint that my definition is too narrow on the other side, that on it 'mystified members of the working class' could not be said to have an ideology. But I can easily meet this by saying that they have taken over, or have had imposed on them, the ideology of the ruling class or of a social order hostile to their real interests.

Secondly there is the charge that I speak as if I had a 'neutral standpoint', but that what I have assigned to this rôle is just one ideology among others, namely 'British common sense'. I do indeed think that there is an absolute distinction between truth and falsehood, and that the definition of 'ideology' must be based partly on that distinction. But so too does Hollis, when he refers to groups which are primary in a *true* social theory, and again to *misunderstanding* and *mystification*. Equally I think that there is a neutral way of getting towards the truth, which might be called common sense or scientific method or the principles of confirmation, and which has sometimes been called reason. But this is something which men of all cultures for many thousands of years have been able to apply well enough to at least some of their concerns, however hard it may be for philosophers of science to formulate it precisely or to justify it in the face of scepticism. In arguing that some part of some ideology is distorted, we appeal to kinds of evidence on which the adherents of the ideology themselves rely elsewhere. I neither need to nor do claim any peculiar possession of this 'common sense', and it is only some kind of inverted racialism that would label it 'British' with a sneer. No doubt we all suffer from some degree of conceptual imprisonment, but there are well-known ways of working towards release. It is not being insular or arrogant to suggest that beliefs and theories in all fields should be tested by seeing whether or how often their distinctive predictions are fulfilled, or that one should be suspicious of *ad hoc* remedies when they are not. Of course *A calls* magic what *B* may *call* technology, and vice versa; it does not follow that both views must be equally defensible. Hollis is trying to score a point here by suggesting the denial of what he repeatedly concedes elsewhere.

Hollis gives, in the passage starting 'The problem posed is familiar . . .' an admirable picture of the line of thought by which so many social theorists have come to assert the relativity of truth. But a little critical reflection would show that the slide is quite unnecessary, that it trades upon a cluster of confusions. It was one mistake to start with anything as extreme as 'Positivist criteria'; another, partly consequential, mistake to regard other cultures as wholly 'pre-scientific'; it would not be so easy to show that 'common sense' is itself unscientific by its own criteria; the contrast between 'relational' and 'relative' seems to be a distinction without a difference; and of course an issue of truth and falsehood arises only with respect to a definite *proposition*: it is obvious that the same *formula* in different languages may represent different propositions which may have either the same or different truth values. Supported in any such way as this 'the extreme doctrine of the sociology of knowledge' does not need to be taken seriously, and it is quite plain that Hollis does not take it seriously himself. Why then does he play games with it to make spurious difficulties for my account?

It is a pity to have to swat butterflies, but they lay eggs that turn into caterpillars.

Index

Achinstein, P., *Law and Explanation*, 5, 16, 30
agent teleology, 95
Alston, W. P., 'The Place of Explanation of Particular Facts in Science', 119
Anscombe, G. E., 77, 88
Aquinas, 82–3, 85, 93
Aristotle, 82, 85ff, 91, 93, 202
asymmetry of causal relevance, 131, 134, 152, 169–71

Barker, S., 45
Berkeley, 136
Braithwaite, R. B., *Scientific Explanation*, 121, 162
Bromberger, S., 'An Approach to Explanation', 30

Carnap, R., 91, 163
categorical framework, 206, 209–11
causal process, 127ff; and closed and quasi-closed systems, 176ff; and pseudo-process, 129ff, 149, 173
causal relation, 126ff; and the logical connection argument, 187, 191ff.
causal relevance, 131, 134, 141. *See also* asymmetry of, 'mark' *and* statistical relevance
common cause principle, 121ff, 138, 165, 167, 171ff
Cummins, R., 45

Davidson, D., 'Action, Reasons and Causes' 12, 13; 'Causal Relations', 10; 'The Individuation of Events', 8
deductive-nomological model of explanation, 4, 10, 34ff, 46, 51, 92, 121, 126, 145, 153, 158–9, 166, 168
defeasibility, *see* efficient causality
Descartes, 92, 107
determinism, 133, 180–4
Dickens, 204–5
Dretske, F., 'Contrastive Statements', 17

emphasis-description defined, 23–4
ends, 76ff
efficient causality, its defeasibility, 92ff, 109–10, 116ff; and tendency, 93ff, 104–5, 106–8
Euclid, 92
events, identity of, 8ff; how identified and explained, 150–1
'explain', meaning of, 47, 57, 66ff
explanation, not an argument, 118ff, 145, 162ff; and causal relevance, 120ff, 152ff; formal problem of, 46; in literature, 204–5; of particular events, 119–20; in social field, 186, 203; unitary concept of, 202ff

Farrer, A., 94
Frege, 80, 92
Fries, 110
'for the sake of', 82–3
free-choice, 89
Freud, 205, 207

Goldman, A. I., *A Theory of Human Action*, 8, 60, 61
Good, I. J., 'A Causal Calculus', 179
Greeno, J. G., 'Explanation and Information', 119, 163
Grunbaüm, A., 128, 167

Harvey, 91–2, 107, 115
Hempel, C. G., 51, 72, 119, 152; *Aspects of Scientific Explanation*, 6, 10, 11, 121, 155, 159, 162, 163; 'Explanation in Science and in History', 118
Hertz, 135
historical explanation, 203–4
Hume, 128, 134, 139, 140, 144, 148, 212
hypothetico-deductive method, 92–3

ideological explanation, 186ff, 198, 206, 207ff, 211–14; and causal explanation, 187ff, 196–7, 201ff
ideology, definition of, 185, 195, 198, 206, 211–14; and intelligibility of actions, 190, 193; involves falsehoods, 185, 194–6, 199–201, 208; and objectivity, 208ff; and scepticism, 196; in thinking about society, 186, 198; and values, 195–6;
inductive logic, 183
Instrumentalism, 136, 138, 144
intentions, 85–6, 101–3, 108, 112–13, 191ff.

Jeffrey, R. C., 'Statistical Explanation vs. Statistical Inference', 119–20, 162, 183

Kant, 110
Kenny, A., 87, 89, 91, 104
Kim, J., 'Events and their Description', 21; 'On the Psycho-Physical Identity Theory', 8,
Kneale, W., 91
Körner, S., *Categorial Frameworks*, 209–11
Kotarbiński, 76

Lenin, V., 188
Lewis, C. I., 136
Locke, 140

Mach, E., 136
Machiavelli, 205

Mannheim, K., 185
'mark', 129, 139ff, 146, 148ff, 169ff. *See also* Reichenbach
Marx, 185–6, 188, 190, 200, 205
McTaggart, J. M. E., 93
micro-explanation, *see* self-explanation
Mill, 85, 93, 131, 156, 172, 200
Moore, G. E., *Principia Ethica*, 91

Nagel, E., *The Structure of Science*, 6, 121, 162
Nelson, L., 110
Newton, 92

object of explanation, as event, 7–9; as linguistic entity, 10ff, 48ff, 53, 73–4; as non-linguistic, 6ff; as non-linguistic under a description, 10ff, 49; no object theory, 37ff, 63–5, 75; as a question, 28ff, 74; as state of affairs, 9–10, 54ff
objectivity, 208ff
Oppenheim, P., 6, 10, 119
Orwell, G., 205

Paley, 117
Pareto, 185
Pascal, *Provincial Letters*, 114
Phenomenalism, 136, 138
Physics and geometrical reasoning, 92ff, 116
Plamenatz, J., *Ideology*, 189
Plato, 81, 205
Poincaré, *La Science et l'hypothèse*, 116
Popper, Sir, K., 142, 162; *The Open Society and its Enemies*, 187
Positivism, 208–9, 216
practical reason, compared with theoretical, 88ff, 109–10, 115–16. *See also* teleology
Proust, 205
psychological egoism, 79, 97–8, 113
purely teleological propositions, 80ff, 95; their limited extensionality, 83–4, 101–2
purposive explanation, 191

quantum mechanics and the violation of causality, 133f
Quine, W. van O., 77, 79

Reichenbach, H., his basic principle of explanation, 124; *The Direction of Time*, 122, 129, 131, 133–4, 139ff, 142, 146, 152, 165, 167, 169ff, 172, 179; his argument for theoretical entities, 136ff; *Foundations of Quantum Mechanics*, 133; on 'unreal sequences', 130–1, 173; on the screening-off relation, 134, 142ff, 156
Russell, B., *The Analysis of Matter*, 177; *Human Knowledge; its Scope and Limits*, 122, 128–9, 134–5, 139–40, 144–5, 165, 174; *Principia Mathematica*, 92

Salmon, W., *Statistical Explanation and Statistical Relevance*, 118–19, 122–3, 126, 134, 162, 164; 'Russell on Scientific Inference', 145
Scepticism, 210
Scheffler, I., *The Anatomy of Inquiry*, 10, 39–40
School of Critical Philosophy, 110
Scientific Realism, 144–5
'screening-off' relation, 134, 142ff, 153, 156–7
self-explanation, 9–10, 14, 49–50, 60ff, 72–3
sentences, basic descriptive, defined, 20; basic interrogative defined, 22; preferred, 5; restructured 1ff.

Shakespeare, 204
social existence, 186; and its relation to consciousness, 187ff
sociology of knowledge, 185, 195–6, 198, 207, 210–11, 216
Sorel, 185
spatio-temporal continuity, 127–9, 132–4, 141ff, 157
Spinoza, 93, 205
statistical explanation, and the S–R model, 118ff
statistical independence of events, 122ff
statistical relevance, 126ff, 141ff, 146ff, 178ff
Stoutland, 191
Suppes, P., *A Probabilistic Theory of Causality*, 179
Swift, 204–5
symmetry of prediction and explanation, 35–7, 52–3, 74–5

teleology, and ends, 86–7; in biology, 87, 104, 107, 111–12; and efficient causation, 85, 93ff, 104; and extensional logic, 80, 83–4, 94ff; and the logic of practical reason, 87ff, 116–17, 191; and the science of values, 110. *See also* ends, intentions, practical reason *and* wants
theoretical entities, 127, 135ff

Utopian Socialism, 188

von Wright, G. H., 191, 193

wants, 77–9, 95ff, 112ff, 192, 201
Wilson, G., 37, 45
Wittgenstein, 78, 84
Woodger, J. H., 92